Wild Hog (see pg. 120)

Doane

Lore of
The California Vaquero

It was the Bridle Which Saved José Juan's Life (pg. 58)

Lore of The California Vaquero

by

Arnold R. Rojas

This book is a complete, unabridged edition of the original classic by Arnold R. Rojas (1896-1988). We have kept it exactly as the edition he published in 1958. Rojas was 62.

The publisher wishes to thank Mr. Rojas' family, especially Ms. Juanita Montes for help in this project.
We thank you, the reader, for your interest in the work of this important author.

William C. Reynolds, Publisher
Alamar Media, Inc.
Santa Ynez, California
www.oldcowdogs.com

Printed in the United States of America

ISBN: 978-0-9890701-0-2

Dedication . . .

This book is dedicated to my friends and neighbors, the people of Bakersfield and Kern County, California.

Foreword . . .

"Bernice Harrell Chipman and I have both read the initial efforts of Mr. Rojas and we believe the stories will be very interesting to a large number of readers. The author has been among vaqueros since he was nine years of age. For some 45 years he has spent most of his life either in the saddle or with horses. They have been his livelihood. He knows the horse without foolish sentiment or illusion, yet with a deep affection and understanding, and well he should, for he has participated in and has seen the passing of the old California style of cattle handling and the new and there was much in the old to interest the student of this phase of our history in the west.

"Mr. Rojas is a true Californian. He was born in this state. His great grandparents came to California from Sonora in the 1820s. He has worked as a cowboy for the Tejon Ranch, Miller and Lux and the Kern County Land Company and he knows whereof he writes when he deals with the oldtime western horseman.

"Mr. Rojas is a self-educated man and he has done a very good job of it. Since he was nine years of age he has been out in the world, making his own way. He has read widely, particularly of the west and its horsemen and he knows the genesis of the Spanish and Arabian horsemanship and its influence on the California vaquero.

"We believe he writes very well indeed and with great interest and that many persons will learn more about the old horsemen of California and the San Joaquin through reading his stories. I like to think that my encouragement, in part at least, has been responsible for Mr. Rojas' writing."

Jim Day in the *Bakersfield Californian*

Introduction . . .

Years have gone by since the last vaquero, Spanish horse and longhorn steer passed from the soil of California into the pages of history. Only memories of them remain with a few old men whose ranks are thinned year by year and whose tales of the old days are stilled as, one by one they leave us. In a little while these survivors of a day now gone forever will be gathered to their fathers and posterity could be without a record of the vaquero. Since the children of these men have failed to recognize the historical value of the recollections of the old people, a time could well come when scholars would search in vain for accounts of lost legends, folklore and customs of this colorful period of California's past, the saga of the vaquero.

Encouraged by James Day of the *Bakersfield Californian,* I determined to avail myself of the opportunity he so kindly offered to publish my work in that newspaper. I set myself to gathering all the fragments of vaquero lore which still existed by seeking out the old timers, the survivors of the great cattle herds, and writing down their stories.

No task could have given me more pleasure. From boyhood I have been an ardent student of the vaquero's methods, customs, folklore and legends and have absorbed this lore through listening to the stories these men had to tell. Whenever possible I have mentioned in the text the name of the vaquero who told me the story. In order to maintain the individuality of each old timer, I have set these stories down in the loose and disconnected manner in which they were told, tales of a time on which the shadows of forgetfulness have fallen. If I have at times made errors in paying just tribute to the sturdy men who, by hard riding, helped build the great state of California, it is because *"cada uno es como Dios le hizo y aun peor muchas veces."*

My sincere gratitude goes to the old vaqueros whose stories made this book possible, to Mr. and Mrs. Adam Mulhause for pictures of the old timers, to genial Frank Bogart of Palm Springs for pictures of antique bits and spurs in his collection, to Frank Lane of Lancaster for help in collecting material, to Amado Guzman of Tepic, Nayarit, Mexico for the drawings, to Louise A. Renfro for help in preparing the manuscript, to Frank Latta, to the Charros of Old Mexico for help in determining the origin of saddles, bits and spurs and in gathering historical material, to Richard Bailey, the Kern County Historical Society and to Ralph Kreiser for their warm, friendly, encouragement, to Bernice Harrell Chipman, to Walter Kane and to Jim Day for permission to use the stories published in the *Bakersfield Californian.*

Tragedy Avoided . . .

All afternoon Santiago had sat beside me on the creek bank and told stories of another land and another day; tales of hard fights and quick victories, of sudden defeats and desperate flights to escape death. The following is the story I liked best.

"All day we had been running for our lives, with the men who wanted to kill us in close pursuit. Their horses had been fresher than ours at the start of the chase, but those who pursue are never as desperate as those who are pursued, so we had kept far enough ahead to be out of rifle range, though by the time we reached our first cover in all that long hard ride our horses were sobbing for breath.

"It was a deep wooded canyon where we dismounted, abandoned the horses, crawled into the sagebrush and willow thickets away from the men who wanted our lives. Our pursuers stopped when they reached our panting mounts, caught them and drew back a distance, for we had guns and could shoot from cover. But we did not feel like fighting; we were only too glad to make our escape. After hours of inching our way through brush, over rocks and cactus, we came to a small open flat between the creek and foothills and stopped to gaze at a drama being enacted before us.

"'There in the middle of the flat were ten or twelve half starved woebegone mares. They stood in a circle, facing the outside where four wolves and three coyotes were circling them. Inside the ring formed by the mothers were a number of mule colts, their ears forward and their nostrils distended in terror, standing quietly. Their instinct told them this was the safest place. The mares had their necks stretched toward the wolves, their ears laid flat against the heads and their lips were pulled back from their bared teeth. They were shaking their heads up and down in vain, futile menace.

"We stopped at the edge of the brush as the wolves trotted around them, seeking an opening. The wolves decided on an old white mare as their easiest prey and one made a leap for her. She struck at him but he evaded her forefeet, slashed her shoulder and blood streamed down her legs. We dared not shoot because the men who wanted our lives were still in the vicinity, nor did we dare leave the cover of brush which hid us. So we sat there, powerless to help and waited for tragedy to overtake those poor, pitiful mothers.

"The wolves grouped in front of the old mare and rushed at her. They had pulled her to her knees when we heard a loud bray and then approaching hoofbeats. Charging around the hill with head along the ground, ears laid back, teeth bared, mouth wide open and braying in a high pitched scream came a *burro manadero,* breeding jack, and fell upon the wolves. They ran off over a hill and we saw them no more.

"No cavalry troop with guidons flying and sabres flashing could have been more dashing than this braying, biting, striking burro protecting his mares and colts. After a while he returned from chasing the wolves, the mares broke their circle and resumed licking the scant grass while the mule colts suckled contentedly. Danger was over.

"That night we crawled out and cut up part of the mare the wolves had pulled down and when we were sure our pursuers had gone back to the town where we had fought, we made a fire, roasted and ate the meat."

Rattlesnake . . .

Don Leonardo went to bed every evening at dusk and got up every morning at four, summer and winter, a habit formed in his youth and one which he saw no reason to break. He dried his apricots, prunes, tomatoes and sliced pumpkin, *orejones de calabasa,* and made his own wine which was stored in a shed adjoining the barn. Among the things kept in this shed was a roll of bedding used on pack trips and a bed spring which lay half buried in straw.

I had arrived by a late train and had walked the short mile from the depot. It was too late to go to bed in the house without awakening the old man so I went into the shed, spread the bedding on the spring and lay down. After I had lain quietly for a time the night noises began. I heard the sound of something crawling and a little later, a rat squealed. There were many gopher snakes about the place but we were careful not to hurt them. They are valuable to the rancher since they kill mice and rats. For this reason they were forgiven if they sometimes ate a chicken.

It was very dark and I felt that something evil menaced me in that shed and that I was in grave danger from some crawling thing. I was afraid to go to sleep and I thought of sweeping my arm around to feel for whatever was there. Luckily for me I did not, but lay there half afraid to move. Even when the noises had all ceased I was still uneasy and at last lit a match, sat up to look around, and found myself staring into the beady eyes of a coiled rattlesnake. It was asleep because it did not rattle or strike, just stared. Snakes don't have eyelids and sleep with their eyes open. I have always believed that I made it in one jump from a sitting position on the bed to the top of a barrel along the wall. Stepping from one barrel to another I reached the door, opened it and stepped to the ground. I knew that a lantern hung in a shop near by so I got it, lit it and hung it on the tines of a pitchfork. I stood outside the door and with the fork held the light over the bed but my movements had awakened the snake and it was gone.

I wouldn't go back into the shed but climbed into a buggy and sat there until I heard Don Leonardo stirring in the kitchen. I at once told him of my adventure with the snake. He gave a start and asked, "Are you sure you were not dreaming?" I led him to the shed and showed him the place where the snake had coiled. We looked around and found it asleep under a corner of the barn. We killed it with a pitchfork. Don Leonardo and Dolores both said, very seriously, that God had been watching over me that night. This rattler had thirteen rattles and a button. I cut its belly open and found a big rat which it had swallowed. That probably accounted for its logginess and failure to strike at me when I lit the match.

People say that each rattle represents a year in a snake's age but I have never known anyone who kept a snake long enough to determine the truth of this belief. One day Ray Foster and I were riding near Temblor and came upon a full grown snake which had no rattles at all but had a blunt tail. It probably had lost them in an accident.

I have never forgotten the scare I received that night. For weeks following my experience I would awaken at night and light matches to see if there were any snakes in my bed though I slept on a cot in an upstairs room. All this happened in 1914 but memory plays queer tricks for every now and then through the years since, I have awakened at night and for an instant the thought that a rattlesnake is near will recur and give me a start before I am fully awake. During the years I worked on cattle ranches, whenever I slept on the ground, in abandoned shacks or wherever I found shelter I would lie awake, ill at ease because of snakes.

One day my uncle Jim McFall came upon a rattler in the road and it started rattling. He picked up a rock and hit the snake. Believing he had broken its back and that it could not strike, he reached down to pick up the rock again and the snake struck him on the back of the hand. He made a tourniquet of his bandana to stop the circulation and finished killing the snake. He cut two slits crosswise in the bite, cut the snake open, extracted the gall and applied it to the wound. For hours afterward he would suck the wound, spit out the blood then rinse his mouth with whiskey. Whether the gall drew out the poison or whether the snake's supply of venom was too low to kill him (which is sometimes the case) I don't know, but the old man survived the bite and lived to kill more snakes. There were many of them about.

Whenever he could kill one in its sleep or before it coiled and struck itself, he would skin and dress it, cut 6 inches from each end, put the fat in a pan and set it on the roof to be rendered by the sun. The rest which is all vertebrae he would hang from a limb. After the meat was jerked we kids would break off a piece and eat it. In time the jerked snake would be gone but we suffered no ill effects from eating it, for we were young!

The Prince Albert Coat . . .

Whenever the vaquero returned to the ranch after a trip to town he always brought back some article he had purchased while his money lasted, if nothing more than an harmonica or a pair of levis. The other vaqueros never failed to pass judgment on the new possessions, either approving or disapproving of his wisdom or taste.

I had brought back a new bit, a pair of boots and a jacket. Vicente, after inspecting my purchases, said, "While it is true that the sound of the cricket decides the worth of the bit and that by spinning it the horse indicates his contentment and this movement prevents the mouth from becoming dry and parched, it is also true that crickets and creaking boots keep one from stealing horses or seeing a girl against her father's wishes. One must tie the cricket and oil his boots before setting out on one of these expeditions."

The blanket lined jacket reminded Vicente of a man who once lived near his village and he told me this story. "Of all human failings," he began, "that of vanity is the worst and sooner or later the vain man will meet with humiliation. Panfilo was the shiest, most modest of the men in our village until the day that he put on a coat for the first time. He had never worn anything more than a blanket with a hole in the center to put his head through. This garment kept him warm on the coldest days.

"One night he was riding home along the river bank and noticed that the water was rising above the usual high water mark. It was well after midnight. The inhabitants of the little village which lay in the lowlands, were all asleep and no light burned in the place. Panfilo went from house to house arousing the people who evacuated their homes before the waters reached them so no lives were lost.

"When the flood had subsided and everyone had moved back into town, the grateful citizens presented Panfilo with a Prince Albert coat, the first ever seen in the village. He donned the garment in the presence of the entire town which had gathered to see the presentation, and immediately his character changed. He who had been so shy and modest became all at once vain and proud. He strutted about town until darkness came and when people could no longer see to admire him, he set out for home. He had left his mule tied to a mesquite and when he approached the animal it snorted and tried to jerk away. Still wearing the coat he tried to mount but the mule would not stand and when he finally got astraddle the animal bucked him off and ran away, dragging the bridle reins. Panfilo picked himself up and set out afoot following the mule. At every house he passed he had to fight off the dogs which attacked and bit him. But he met the crowning indignity when entering his own yard his dogs, four in number, not recognizing him in the Prince Albert, set upon him and would not let up until he removed the outlandish garment and hung it on the fence. The next morning he took the coat off the fence and went to the *milpa* where he put it on the scarecrow and there it remained. Panfilo went back to wearing his blanket again, and to his modest shyness."

Mules . . .

Vicente says that he learned from experience never to change clothes during the time he was breaking a mule as the hybrid will be docile and obedient as long as his master wears the same clothes. The moment the man alters his attire the mule will have to be broken all over again. Vicente had been riding this young mule for some weeks when one Saturday, after a fortnight of docile behavior on the part of the mule, he thought him gentle enough to ride to a dance so he cleaned up and put on fresh clothes. When he walked out to the corral, however, the mule didn't know him. Missing the man's familiar odor, he snorted and shied away. Vicente had a hard time mounting and when he was at last astraddle, the mule lit into bucking and threw him so hard that when he was able to set out in pursuit of the hybrid it was too far away and he had to miss the dance.

Vicente says further that a mule will work cattle better, pull more when hitched to a wagon and go farther and faster under a pack than a horse. Although the *paisano* of old California esteemed the mule for its rugged usefulness, he nevertheless shrank from using the hybrid to pull the wooden plow, for the mule is barren and accursed. His hoofs poison the ground and weeds grow where he has trod. The *Californiano* depended on the wide horned ox to till the soil because the ox is blessed and his great hoofs enrich the soil and prune the plants so that the harvest may be rich and abundant.

I asked Vicente why, of the two faithful animals, one was so highly esteemed while the other was so unjustly maligned. He said that it had been thus since that cold night in the stable in Bethlehem when the Babe lay shivering on the straw and the ox approached the manger and warmed the Savior by blowing his breath upon him, while the mule shied away from Him who had come to save mankind.

The Fugitive . . .

The idea of guns and gunplay has always been distasteful to the California rider who shrank from the use of them. The characters in the old tales seem to have adjusted the differences that arose without resorting to firearms. When one found, now and then, some old *maldito* who always carried a gun, he was some old scorpion who without a doubt had been a bandit. The old tales contained little of the blood and thunder one finds in modern sensational literature.

To speak of guns and shooting was to embarrass the vaquero. He knew little about them, looked askance at the man who carried one, and deemed him lacking in manhood. To carry a weapon was to invite scorn and few were ever worn.

According to the old men, the acme of daring was for some young bravo to ride into the middle of the main street of a pueblo, riata in hand, and dare the officials to arrest him. The officers knew that a touch of the spur or pressure of the knees would bring the horse in a leap upon them before they could use a weapon to good effect, so they seldom accepted the challenge.

The following is one of the few tales of California's yesterdays wherein guns and gunplay are prominent factors.

Juan lived in peace respected by all who knew him. A gentle kindly man, he suffered when an animal was hurt or in pain, for he loved horses and dogs. His one passion was to own as many as he could buy.

To see a race was meat and drink and a fast horse was his delight. He entered his stock in every race he could manage to reach. One time there was a race to be held in Salinas and he went there hoping to enter his horses. On his journey he was followed by a bedraggled, nondescript dog of which he had become very fond. It was this mongrel that was to cause a change in the course of his life and make him the hero of this tale which has been handed down as a saga of a man's struggle for peace.

The dog had growled at a deputy who kicked him.

Slick Fork Single Rig Saddle and Ring Bit

Juan Lopez protested and the outcome of the ensuing quarrel was that the deputy shot the dog. The owner swore someday to even the score. Some days after this incident the deputy's body was found in a well. Juan was suspected of having taken revenge because a number of people remembered the quarrel. The authorities looked for him but he could not be found. He had moved his family and horses to a reservation where he had friends. He found an unfrequented spot and built a cabin into which he moved, resigning himself to the anxious, irregular life of a fugitive.

He devised a code of signals so his wife could warn him when his enemies were about. By hanging certain articles of clothing on the clothes line she could tell him when there was danger, and thus he managed to avoid his hunters.

He had friends who would steer the deputies away from his haunts and by being always on the alert Juan was able to work on the ranches when help was needed and still stay at liberty.

In the early 1900's a new sheriff was elected in Monterey County and his chief deputy wanted to capture Lopez since there was a reward for the killer of the man found in the well. He and a local constable set out to get the fugitive dead or alive. They rode into the country where rumor had it Juan was hiding. They had passed well into the borders of the reservation when a heavy hailstorm hit them. After wandering around half the night they stumbled onto a cabin where they were welcomed by the lone occupant who fed them on tree squirrel which they devoured with great relish even though the game they ate was out of season.

They dried their clothing before the fire which was

the only source of light, and slept before the fire until the storm passed. Their host left at daybreak before they awoke leaving the deputy with but a dim picture of his features. They continued their quest but did not find Lopez and finally turned back to Salinas. When they parted the constable asked the deputy if he knew who their host had been. When that individual said that he did not, the other man told him that they had been guests of Juan Lopez. The deputy swore to return and get his man but the other said he would never try again for to do so would be an abuse of hospitality.

The deputy did return but Juan's wife, ever on the alert, hung out the signal flag and warned her husband. Lopez went out to meet the man for he had shown his face only by the light of the fireplace and had been clean shaven on the night the two men had seen him. He now wore a moustache and a cap and was not recognized by the deputy. The two men greeted each other cordially.

In the ensuing talk the man from the sheriff's office bragged that he was there to capture Lopez and take him to jail. Juan asked how he intended to do that little chore if he found the fugitive. The deputy drew a revolver and said, "This way," and shot at a rock and missed it. Lopez then shot at another farther away and split it in two, and said, "Lopez might do this to you while you do that to him. I am the man you are looking for." He held his gun on the deputy until he had ridden out of range.

The officer returned a third time, but as he entered the vicinity of his last defeat a bullet clipped his ear. A second shot tore the rope off his saddle, and a third bullet went through the cantle. He turned his horse and ran him out of range.

When he arrived back in Salinas he resigned his office,

and a few days later died. Some say it was frustration that caused his death, but the old timers say it was fright. When the next sheriff came into office he did some investigating, cleared Juan of the crime and left him to live in peace again with his horses and his dogs.

Caballo Arisco . . .

The first thing a boy on a cattle ranch must learn is to do his work despite whatever *mañas,* bad habits or vices, the horse he rides may have.

It was during rodeo time when Salvador Carmelo went to work on the Tejon Ranch. It was his first time there though most of the vaqueros were old friends. Among the horses in his string was a sorrel which was *muy arisco,* hard to bridle or head shy. Salvador says that Chico Martinez had bitted him up too soon and as a consequence the horse hated the bit and Chico had been forced to snub him to a post in order to bridle him. A boy, if he remains on the ranch, learns to outsmart such animals because the older men discard them when they can and they are left for the youngsters to learn on.

Vaqueros were loath to leave a ranch because after a year or two a rider had developed a good string and had horses to his liking whereas if he left an old string on one ranch and went to work on another he would get the discards of the older men. Vaqueros often said that it took two years of riding on a ranch before a man had a good string of horses.

The struggle to bridle the sorrel on the mornings when he rode him made Salvador late in getting out of camp and the crew would be far away before he got started. During rodeo time the vaquero starts his day at three o'clock. One morning the boy arose, facing the ordeal of

Salvador Carmelo

bridling the sorrel in the predawn darkness. Dreading the jibes of the men on his being late, Salvador put a *jaquima* on the horse instead of a bit. The animal accepted the noseband meekly enough and the boy was mounted and out of camp with the crew. It was midday before the vaqueros, having fanned out in a circle and brought the cattle to the rodeo ground, were ready to start parting out the beef steers. While the vaqueros posted around the milling herd waited for them to quiet down, Don Jesus the *mayordomo,* rode around the herd. This was his usual practice and he probably did it to see the condition of the horses and also to see if any of the men were missing. When he came to Salvador he stopped and with a puzzled frown stared at the *jaquima* on the old bridle horse. After a moment he said, "Little son, I shall work on this side of the herd, Don Porfirio, the *caporal,* on the other. I want you to keep this position and help me when I bring out a steer." Don Jesus rode into the mass of cattle and soon found a steer that would make beef. He put his horse behind it and after many twists and turns drove it out of the herd where Salvador was posted. Through the long afternoon every steer Don Jesus parted out passed by the boy's position. Salvador knew that Don Jesus expected the sorrel to run away the moment he got him in a run after a cow and was bringing the steers out on that side to teach the boy a lesson. But the young vaquero stayed on the alert and did his work without getting his horse out of a walk.

When the rodeo was over and the cattle were left to drift away Salvador had gone through the whole operation without having his horse run away with him. Don Jesus rode up with a gleam of humor in his eyes and said, "Little son, I wanted to see you run that horse but you

George Brunk

were too smart for me." The boy grinned and answered, "Yes, Don Jesus, I know that a *jaquima* does not go on an old bridle horse because it will not control him but that sorrel is, as you know, *muy arisco,* and that was the only way I could get out of camp with the rest of the men."

Bucking Horses . . .

When a man says he has never been bucked off a horse one may be sure that he lies or that his riding experience has been limited. There is no man who has ridden horses any length of time who has not been bucked off. On the other hand there is no horse which cannot be ridden. There are horses which will, through a period of time, throw the best of riders but sooner or later a rider will come along who can handle them.

On all ranches there were to be found horses which were chronic buckers; horses that had been spoiled, distinct from the green colts which were termed *brutos.* These chronic buckers had a certain routine for their performances. For instance, one would buck with the empty saddle every morning and this finished his bucking for the day. Another would not waste his efforts so foolishly but waited until the saddle was occupied before "breaking in two."

Horses bucked for many reasons or for no apparent reason at all. Many would buck on cold mornings as a sort of setting up exercise and to get warmed up for the day. A lot of them bucked when they felt their rider to be off balance. Most of them bucked when something went wrong such as having the rope get under their tail or between their legs. Some bucked once in their lives and never bucked again such as the appaloosa on the

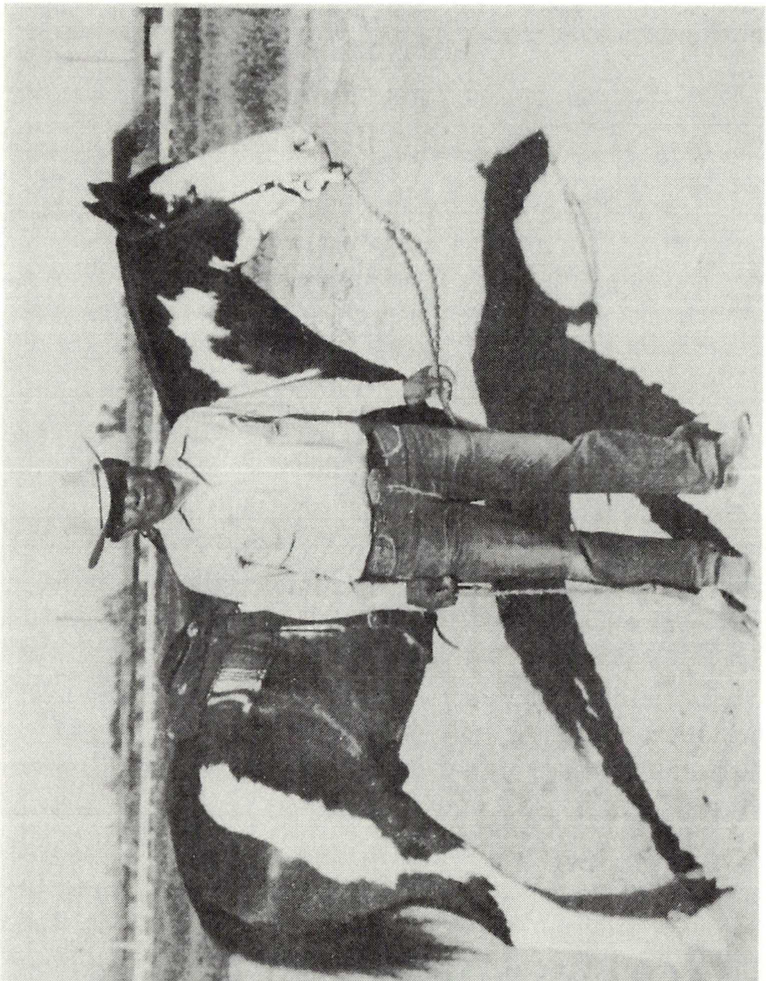

Don James

Rosedal Ranch; he bucked with a Spaniard for so long the man was bleeding. I have forgotten his name but Antonio Moss was there at the time. Others started bucking and became outlaws after years of service under the saddle. Some horses chose one man in particular for their animosity and were well behaved with others. Occasionally there were men for whom a horse never bucked. Some colts would walk off like old saddle horses the first time they were ridden and not even hump their backs until later. When they were bridle wise and half broke they would suddenly "break in two." Other colts bucked the first two or three times they were ridden, then quit for good.

Sometimes a man would have ridden a horse for a season or two, then some fine morning the horse would "bog his head" with his nonplussed rider, the element of surprise being in the horse's favor, the vaquero invariably got bucked off. As a rule colts did not buck hard the first time they were ridden and if the rider stayed aboard they would soon become discouraged and quit. But if they dumped him once they would try again and it is needless to say that they never failed to improve in technique, trying a new trick each time they bucked. Habitual buckers were seldom cured of their bad habit and they would try to throw their riders every time the condition arose that prompted them to buck the first time.

Though there were horses on every ranch which were distinctly dangerous there were also men who could handle them. The horses which bucked were no more vicious than the others. They had simply learned bad habits. The instinct for bucking is not confined to one particular breed. Roland Hill's band of Morgans produced some first class buckers and in some instances the horses of coach blood bred by Miller and Lux made bad outlaws. Among the Thoroughbreds raised by the Kern County Land Company were some individuals which could go "high, wide and handsome." One day on San Emidio, Telephone, a lanky sorrel, bucked a number of times in one morning with George Hoskings then quit and never bucked again. The Tejon Arabs never took a back seat either. A grey *sun-fished* with Luis Zamora and picked up dirt with the stirrups he was so crooked. A bay Arab bucked spinning in a circle when Luis took down his reata to rope calves in the corral.

Good riders of bucking horses did not come from one particular part of the country. There were good men wherever cattle was raised though the best who came here in the early days were from Oregon and Nevada. More often than not, these men were big, husky six-footers. A man's physical makeup, however, had nothing to do with his ability. If a man had the knack he was a good bucking horse rider regardless of size. I remember a little redhead from Kansas who was a good rider, also Mariano Sanchez, tall and thin, a typical Yaqui. Luis Zamora was a small man as were Juan Olivera and Coyote Castro. John Driggs from Nevada was big and husky. Buster Clark of Bakersfield is a big man. Archie Reed from Honolulu is not a big man and George Fields, who came from Oregon and always wore woolly *chaparreras,* weighed over 200. This can be said of all the men who were good bucking horse riders. They didn't *give a hang* if a horse was bad or not.

One year Miller and Lux rented all the saddle horses and the cattle on Carriso Plains to a movie company to make a picture. The company brought its own riders, all rodeo stars. When the picture was finished all the outlaws had been tamed and no longer bucked, though some were nine years old and still in the hackamore. Max Gomez who watched the filming, said that when one man was bucked off another got on until there wasn't a horse that would buck. One little black horse which had bucked his riders off a number of times the year before was being ridden by a girl after the first few days.

Sometimes funny things happened on the ranches. A lad came to San Emidio and got on the vaquero crew. He had heard that if a horse's head was held up he could not buck. So, when Henry Hoskings put him on old Gacho, the boy, instead of letting the reins loose and letting the horse walk off, held him tight. The bewildered Gacho backed a few steps and finding that the rein didn't slacken, reared, lunged and crow hopped across the yard. The lad bounced out of the saddle over the fork and with both feet still in the stirrups, grasped the bit in both hands and rode Gacho across the yard, sitting on the horse's neck. Cleve Helm and George Hoskings stopped the horse and helped the boy back into the saddle and Cleve christened him "Booger Red" after a famous bronc rider. Booger Red he remained despite his vehement protests, through the years he spent on San Emidio.

Sometimes a horse which seems tame will do the unexpected. Back in the thirties, the Indians from Tejon came into Bakersfield to ride in a parade. We had borrowed mounts for the affair from Earl and Cecil Pascoe. All the Indians were seasoned vaqueros with the exception of two teen age boys and each man was given a horse according to his ability. Two old mares were assigned to the boys, Catarino and Canuto. Each man was issued a costume which included a fine eagle feather war bonnet. They had all dressed, mounted and were leaving the fair grounds to form in the parade when a wind came up. It blew Canuto's war bonnet up and then down over his face which scared the mare. She lit into bucking but Canuto, with a rein in each hand, weathered the storm. As he was wearing moccasins his feet went through the stirrups half way to his knees. Bob Addington and Augustin Hinio, who caught and stopped the mare said that she went so high that they thought she was flying, which proves that one can be deceived in the appearances of either horses or young *paisanos.* When the boys went back to Tejon they told of their experiences in town but the main topic of conversation was the *rodilla* Canuto put on the old mare. To ride a bucking horse was *echar una rodilla.*

The event of a horse bucking with his rider never failed to bring roars of laughter from the rest of the crew. The rider's predicament and his efforts to stay on the hurricane deck struck a humorous chord in the vaquero's makeup.

Among My Souvenirs . . .

Among my "forty years gathering," as buckaroos call the accumulation of a lifetime, is a gunny sack in which I have stored a number of articles which were used by old time vaqueros of this county. Most of these bits of equipment were made by hand and are remnants of their outfits. Too service worn to be of any monetary value, these battered keepsakes are worthy of appreciation because of the associations connected with them. Each one retains some characteristic of its former owner or maker and recalls some old *compañero* or incident of days gone by. Vaqueros were men in love with their craft and they clung to these souvenirs of a calling that has passed into the limbo of time; mementos of their sturdy, dauntless youth. To add to my collection I have had to wheedle every one of these dilapidated links with the past from the old timers.

In the old days a buckaroo, when applying for a job would spread his outfit on the ground for the boss' inspection. Remembering how well an old time foreman could judge an applicant by the condition and quality of his equipment, I emptied the sack out on the ground to see what Paul's opinion would be in respect to the tack. Paul Apodaca is white haired, slight of body and frail looking, soft voiced and shy in manner. He is Pablito to his intimates. His appearance is misleading, however, for through the years few men have earned so wide a reputation as a vaquero. His years of experience with horses have taught him the value of infinite patience in coaxing a refractory colt into good habits and submission. From the quiet way he goes about handling stock one can hardly imagine him as the lad he was years ago. In looking over the equipment on the ground his expert eye at once recognized the make or style of hand work. The slick fork Visalia saddle ridden by Ed Valdez for thirty years on Tejon Ranch he knew of old and said it was still good. The Spanish bit, perfectly balanced and beautifully inlaid with silver was the work of A. B. Hunt, considered the finest bit and spur craftsman in these parts in a day when a vaquero would never have dreamed of using bits or spurs that were not hand tooled to order. Mr. Hunt, who in his youth was a good vaquero, still does a little of the skilled work for which he was noted. His brother, George Hunt, was foreman on Tejon for many years.

The rawhide reata was braided by Pete Villareal on Tejon Rancheria and the old *jaquima* was the work of

Salvador Carmelo, San Emidio vaquero. The hair rope with a center or "heart" in it was made by Lupe Valenzuela who made the best ones hereabouts. All good hair ropes have "hearts."

The work on the quirt was peculiar to the Felizes. Almost all the vaqueros on Lakeside Ranch were members of this family, Joaquín Feliz being foreman there for a long time. His nephew Frankie made the quirt. The spurs with the rowels worn paper thin had belonged to old Jim Loveall at San Emidio Ranch. Old Jim was one of the most self-possessed men I ever knew. One day a horse he was riding fell with him, smashing the left stirrup and binding his foot. The saddle was pretty loose and Jim knew that if the horse rose to its feet the animal would drag him. Cool headed old Jim reached over, caught the bit in his right hand and twisted the horse's head around so that it could not rise, then, changing hands he uncinched the saddle on the off side and when everything was clear, let the horse up. By the time we got there to help him he was prying his foot out of the broken stirrup and mumbling something about horses that couldn't keep their feet under them.

One of the two chains on each spur is longer than the other and for a purpose. With a slight movement of the foot the chain will hit the iron cover of the stirrup and signal the horse into instant alertness. Then too, a horse will keep pace with the tap, tap of the chain against the stirrup and thus the vaquero will have rhythm to brighten his day.

The particular style of braiding and the buttons on the rawhide reins were strange to Paul since they were not made in this part of the country but over on the coast. The *fiador* and *tapa-ojos,* blinds, had belonged to Frank

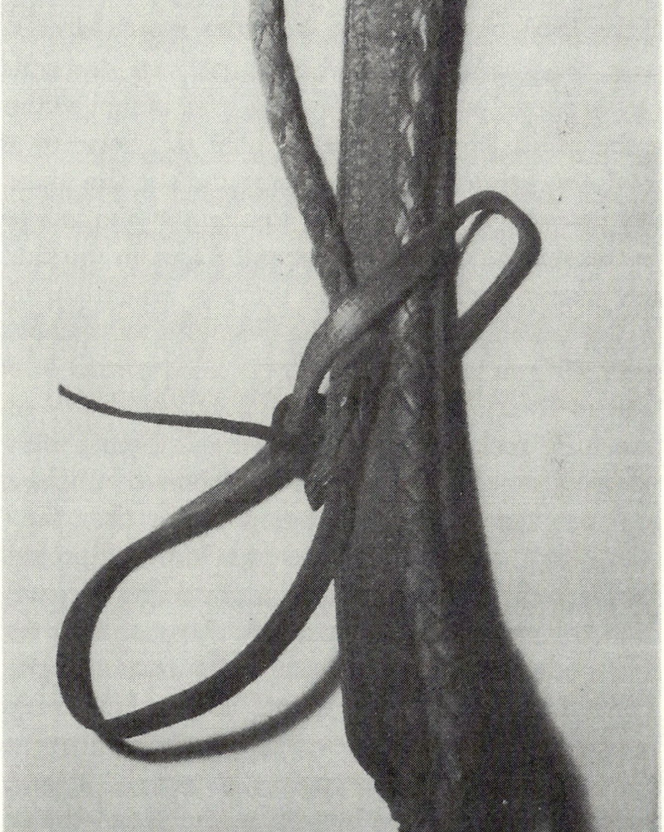

Tapaojos

Urrea, Poso Ranch foreman in the early twenties. Paul said the *tapa-ojos* reminded him of a wild ride he had on a bronco when he was a boy on Tejon Ranch. His brother, Frank Martinez, a good deal older than he, often counseled him and said that certain methods were used for certain reasons and that the old people had found through experience the old ways to be the best but all this fell on deaf ears. In those days blinds were used to cover the bronco's eyes when his rider mounted or dismounted. Paul, who was breaking a colt, one day undertook to improve on the old system and mount without using the blind. He first tried to hold the horse by the cheek of the bridle while he swung on but the animal, a husky fellow, was too wild. He finally tied him to a post and in this way he was able to mount. As he hit the saddle the colt reared back, pulled up the post and stampeded across the country dragging the post like an anchor at his heels. He ran in a circle with his head turned to one side and in frantic terror of the attachment. Every time the post hit a rock, and there were many around the vaquero camp on the Tejon, it would bounce up barely missing the rider who was hanging on for dear life. If he had fallen the post would have hit him and probably broken his neck. A long time and many miles later when the bronco had begun to slow down, the post hit a rock, bounced up and dropped in front of Paul who caught it and headed the tired horse home. The animal, played out, ignored the added burden. His adventure brought a laugh from the men but he narrowly escaped a licking from his brother who gave him a long lecture on the evils of trying new fangled ways. Paul says that after that lesson the horse was broken to mount without blinds and to carry or drag anything.

These souvenirs of another day were important to the vaquero who became attached to them because his life depended upon them and they are valuable to me for their associations with men whom I knew and greatly admired. To me that gunny sack holds the ghost of each old vaquero's buoyant, irrepressible youth.

The Amansador . . .

As a boy Juan Ardia suffered much privation and want. His father, who had been a good provider, died when Juan was a small boy and an uncle took him into his home. This uncle was a poor man and when the meager supply of food he was able to supply his family was shared among its members, Juan's portion would be a spoonful of *garbanzos* and a tortilla or two.

On the first morning in his new home his uncle awakened him at three o'clock and asked the boy if he was sick. Juan asked why and the uncle answered, "It is after three and past the time for milking the cows."

Juan did not have meat to eat and was always hungry. One day he was ordered to sharpen the knives. He obeyed cheerfully because there were some fat calves in the corral and he thought his uncle planned to butcher them and he would feast on beef. He was disappointed for the next morning his uncle used the knives to earmark and alter the calves. It was only after he learned to ride bucking horses that he began to eat meat.

People brought him their colts to break, and in time he got the reputation of being able to get the bad habits out of a horse when others had failed. For as Juan says, "Anyone with patience can in time teach good habits to a horse that is not spoiled, but a man who can take one that is spoiled and cure him of his bad habits is a much superior horseman."

One day they brought him a horse which no one could stop or turn. His jaws were numb and the bars of his mouth were dead and without feeling. He did not want to take the animal and told the owner so but that individual told Juan that his rivals, the other *amansadores* of the neighborhood, had given the horse up because they didn't have the nerve and the know how, so Juan finally accepted the task. His first act was to sharpen his knife to a razor edge. He then took the horse and tied him securely so that he could not struggle and hurt himself. With the sharp knife he cut the calluses off the bars of the horse's mouth. After this operation he put the animal in the corral and fed him tender grass until his mouth was healed. When the horse was sound again he began to bit him up. Day after day he put the bit with the copper rings in his mouth after sprinkling it with salt to make it more agreeable and to help harden the tender spots.

Time passed, for people were never in a hurry in those days. The horse, like the prisoner who missed his chains after they were removed, made a habit of having the bit in his mouth and finally, as they say, "took it." With the bars of his mouth no longer callused and dead to feeling but new grown and sensitive, the horse obeyed the pull on the rein and again stopped and turned as he did before he had been spoiled.

On another occasion a horse was brought to Juan. He opened its mouth to see how much harm had been done and found that the tongue had been cut half in two and there was a cavity in one of the bars from pressure of the bit. He drew a design for a bit and went to his bit maker with it. The man was puzzled and asked him if he was sure he wanted it that way as it was a most peculiar looking affair. The bar on one side of the port was three quarters of an inch higher than the other side. This was to avoid the hurt place in the horse's mouth. Being relieved of the pain which had made him frantic the horse no longer ran away but became a useful animal again.

Cattle Drives . . .

Virginia City, Nevada, during its hectic, booming days was a sales point for California beef which was a drug on the home market during those times. Cattle moved in a steady stream from Southern California and Coastal ranches along the edge of the desert, following the eastern base of the Sierras to the bonanza towns of Nevada.

Vaqueros, born and reared in the more temperate climate to the south, returned home to tell unbelievable tales of blizzards and storms, of forced marches across arid stretches to the bleak, mushrooming mining towns. San Luis Obispo, Tehachapi and Los Angeles were starting points for these drives.

Now and then while traveling through some remote town in Nevada, one hears a familiar Spanish name and on inquiring finds that some vaquero who came with the trail herds from California had been persuaded to stay and take root here among strangers.

I once mentioned to Perfecto Cuen that I had met a man named Ledesma in a little town in Nevada and Don Perfecto said, "I knew him. We were in the same vaquero crew when we drove a herd from here into that state. When we left there he stayed behind and never came back to California. He is very old now because we made that drive near the middle of the last century."

Tom Eveleth Sr., one time head of the revenue collector's office in Bakersfield, spent his working hours at a desk but his thoughts may have occasionally wandered to his early life on the ranch to recall many of his experiences.

Tom has bought many head of cattle in his lifetime and he is one of the best judges of beef in the country.

Whenever we meet for a chat our conversation always turns to cattle, vaqueros and good vaquero horses. The other day he told me of an incident which occurred on a drive many years ago. Tom Briggs was in charge of the Miller and Lux buckaroos when they started out of Santa Maria with a herd of poor cattle. They drove slowly the first day out and made so little progress that it was long after dark when they made camp. The cattle laid down immediately and Briggs detailed two vaqueros to guard them.

About midnight the moon came up and one of the vaqueros, a boy, woke and sat up to discover that he was sleeping in a graveyard. He jumped up with his bed canvas draped around him and lit out of there as if the devil were behind him. The sight of the billowing canvas coming toward them scared the cattle and they stampeded off across country. Tom says the animals were so weak they couldn't run more than four or five miles so the crew had little trouble in gathering them the next morning. The kid gave them more trouble than the cattle. They couldn't get him to bed down again that night.

Leonardo Chavarria . . .

During his boyhood Leonardo Chavarria lived on the Mojave Desert near La Laguna de Chico Lopez, now Elizabeth Lake. His father was a vaquero for Francisco "Chico" Lopez.

One day when his mother was alone with the children, two men on jaded horses rode up to the house. The man who acted as spokesman asked for Chavarria's father and on being told that he was absent, asked the mother if she would cook something for the two to eat. The mother said, *"Con mucho gusto"*, and in a short time had a meal ready. The spokesman asked her to set the table in the doorway and holding their horses' lead ropes, the two men sat so that they could keep a wary eye on the back trail.

When they had finished the meal of pounded dried meat, eggs and tortillas, the spokesman asked Señora Chavarria what they owed her. She told them the food was not for sale and that she was happy to give it to them. But the man insisted that since the meal was ordered it must have a price and on departing left $3.00 on the table. The two men then mounted and rode away toward the north. The only thing Chavarria remembers about the appearance of the men was the huge moustache worn by the spokesman. This was the first and only time he ever saw Tiburcio Vasquez and his henchman Procopio.

When Chavarria was old enough to go to work his first job was on the Tejon ranch. He was paid $15.00 a month for the first year and an increase of $5.00 per month each year thereafter until he received $30.00 a month which was top wages in those days. At that time the area of the Tejon Ranch was much greater than it is today. The mesa southwest of Lebec was under the Tejon's control and Chavarria was ordered to report there for work. He was in need of clothes so before he went he borrowed $6.00 from J. J. Lopez the *mayordomo* for that purpose. He recalls that he paid 30c for a pair of trousers and 25c each for his shirts. His entire outfit for the year including boots, came to a few cents less than $6.00.

Russian Drag . . .

The kid was worthless. He couldn't have driven a cow down a lane unless the animal wanted to go that way. The boss had hired him, not for the work that he would get out of him but from a feeling of pity. The boy wasn't cut out for the work of a cowboy and blundered in everything. If he was sent out alone he got lost or dismounted and let his horse get away from him and had to walk back to the ranch or camp. To make matters worse he strapped a gun to his waist and while out on the range shot at every thing that would make a good target, to the great annoyance of the other men who were afraid he would shoot someone by accident because, as they said, "He couldn't have hit the broad side of a barn if he were locked in it and all the windows closed." However, the men let him alone; that is to say they never rebuked him for not doing his share of the work and causing them trouble but ignored him altogether.

One day the boss' horse fell with him and when he got up his foot was caught in the stirrup. The horse started to run but before he could get going the kid drew his gun and shot the horse, killing him instantly. Thereafter every few days the boss would ask the kid if he had plenty of shells for his gun and if not to let him know so that he could buy them for him because he didn't want him to ever run short of ammunition. No doubt he attributed the boy's sudden accurate marksmanship to the hand of Providence.

To have the foot caught in a stirrup when a horse falls or when one is bucked off and to be dragged by the frantic, kicking animal is one of the most terrifying experiences a rider can suffer. In our crew was a man named Bill. He was one of those riders sometimes met with on the range.

He never rode in the dust when the crew was moving cattle for he invariably managed to work his way to the head of the herd, and there he would stay on the point in the nice clean air, and ride along admiring his figure on horseback.

He had another habit which is frowned upon by cattlemen; he was continually turning, stopping and backing his horses to improve their rein until he wore them out and they would do nothing but rear and switch their tails. He spoiled every horse he rode. He may have thought he was doing the right thing but finally the other members of the crew quit speaking to him. Nevertheless, he probably saved my life.

He was just about ready to call for his time when one day we were working a bunch of cattle at El Chicote on the Carriso Plains. The cattle were bunched near a trough by some cottonwood trees and Ray Foster was parting out beef. Bill rode away to fix his saddle, another way he had of loafing, and was sitting some distance from the herd when a steer broke out. I spurred my horse after it and the fast little bay I was riding overtook it where Bill was sitting on his horse.

An animal, when overtaken, will stop and turn back. The horse will stop and turn after it. The little bay stopped to turn after the steer and stepped on a round stick which rolled and he fell. A few days before I had received a pair of Chihuahua spurs with four inch rowels from Porter's in Phoenix. I had filed the steel on each side until the rowel had a half inch play. This made the spurs jingle a lot louder but the main purpose in filing the rowels was to permit them to get crossways and catch, making the rowel into a hook, which, when thrust into the bars of the cinch would hold a rider on a bucking horse. Most old time riders rode their cinch when a horse bucked in spite of the fact that it was a dangerous practice. If a horse were to fall over backward the rider could never get clear. The little bay got up with my left foot in the stirrup and the spur hooked in the cinch. The gentlest horse will be terrified at anything hanging from the stirrup and dragging on the ground and will try to kick himself loose from it. The little bay started to run, kicking at me as he went, but before he had gone far, Bill, the useless Bill, spurred his horse into mine, knocking him down and before he could get up again Bill dismounted, caught the bay's bridle reins, reached over and got a hold of my leg, gave a hard yank and pulled my foot out of my boot.

Bill, the guy who watched his shadow, had more presence of mind than the old vaqueros who watched me go through a similar experience a year later and only a few miles from El Chicote, the scene of the first one. It was late February. The winter had been wet and mild so the grass on Carriso Plains was tall and green when I arrived at the chuck wagon which was stationed at Painted Rock Pump. Kindly old Guadalupe Valenzuela was the cook. I had left Los Angeles wearing a pair of low cut shoes, corduroy balloon pants, this was the era of the Charleston, and a cap. When I appeared before Cuen, the boss, he looked me over from head to foot, grunted, frowned and asked gruffly where my saddle was. I told him it was coming by express but I didn't say when. He knew I didn't own a saddle so he had me take the chuckwagon mules and haul wood until I had a big pile of it for Lupe to burn. On the third morning he told me to take his horse and help the boys move cattle from Painted Rock Pump to Pimentel. We were driving cows and calves through a narrow crossing between the main Soda Lake and a smaller one when a calf broke out. I went after it and the horse, which was a good one, soon overhauled it. It had rained the night before, so when he turned, his feet went out from under him on the slick ground. He fell and one of the low cut shoes I was wearing slipped through the stirrup. The horse arose immediately and ran off sideways, kicking at my dragging form. A horse kicking at a person dragging from a stirrup usually kicks over him and that is what happened in my experience. Only one kick reached me and when it landed in my ribs I thought as I dragged, "He will soon land one on my head and nothing will hurt anymore." But the "Old Man up above had His arms around me," as cowboys say when they have a narrow escape. The horse ran into the lake, mired down and stopped. I pulled my foot out of the stirrup and after I had the horse out of the bog I found my shoe and the tobacco which fell out of my pocket, and rode back to the herd. That happened thirty years ago and I have never worn shoes since, only boots.

The vaqueros were some distance away and all sat and watched the incident which happened so quickly that none of them moved their horses. When I reached them not one of them said a word, just stared at me with an odd expression on their faces. In the crew were three of Twizzleman's riders. Grizzled Dick Kelly is the only one of this trio whose name I recall. In addition there was one of Cooper's riders, a Danish kid, a vaquero from Chimineas ranch, Jesús Garcia on his tough little roan Nevada mustang and Miller and Lux vaqueros. The ones I remember in this group were Jack Valenzuela, Frank Urrea, Ben Stranger, Shorty Stopp, a redhead whose name I have forgotten, Johnney Light the horse wrangler, a man we called Chihuahua and Angel Montana.

Angel lives here in Bakersfield and I often meet him on the street. He likes to remind me of the time I did the *Russian Drag* on the Carriso Plains thirty years ago.

Latigos . . .

The old vaqueros remembered very well and not without good reason every step in the process of tanning the *coreas,* straps, their parents used to punish them with when they were boys.

To make these *coreas,* hides were selected, pegged out and fleshed, then strips were cut the length of the hide. Only the best part of the back was used for this purpose.

Tallow was rendered in a hugh iron pot. When the cracklings were removed cold water was poured into the grease and after this mass cooled and the grease had risen to the top, the water was poured off. The strips of rawhide were each worked vigorously in this tallow. A pit was dug in the ground and the grease covered strips were put in this pit and covered with manure to near the top of the hole and tamped down firmly. Then dirt was used to seal the contents and a flat rock was placed on top to prevent dogs from digging them out.

After fifteen days the rawhides were unearthed and the work of limbering them up fell to the boys of the rancho. The boys took the rawhides to a tree with a branch that extended horizontally from the trunk. A strip was taken from the pile and a large stone was attached to one end, the other end was thrown over the branch and tied to one end of a pole which lay horizontally over an upright about four feet high. It was like a well sweep. The center of this pole lay in a notch of the upright and fitted loosely so that a boy at the end could swing it easily up and down and draw the rawhide back and forth over the rough surface of the branch. But a boy's arms would ache for many hours before enough straps had been made to supply the needs of a ranch. Each strip had to be worked over the branch until every hair was worn off by the friction and it was soft and pliable. After this treatment the rawhide never became hard again even after it had been wet. Somehow the heat process it went through while buried in the ground must have drawn out the natural glue which is in all rawhide.

It was not the aching arms and hours of toil spent in pulling the thongs up and down that made the old men remember so well. The reason for the poignant memories was this. Whenever they had earned a whipping and were called into their father's presence he would say, "Go thou into the room where the *coreas* are stored and select one that pleases thee." Considering the use the father had in mind for the strap not one was pleasing to a boy. There was no way of knowing which one would hurt the least.

Salvador Carmelo . . .

Some years back Captain McKittrick bred and trained polo ponies on his ranch south of Bakersfield and a number of the local vaqueros worked for him. Salvador remembers the day he arrived at the ranch to report for work. He had just opened the gate when a horse, with Earl James aboard and bucking high, wide and handsome, came around the barn and through the gate. Salvador watched the horse buck into the middle of a field where he threw James who hit the ground so hard he plowed a long figure eleven in the Bermuda grass.

If such things happened to Earl James who was a good vaquero and a top rider, *quien sabe* what would come to pass with Salvador who was only a kid. But he soon learned the system of breaking polo ponies. They were not taught to stop and turn like a vaquero horse but to turn in a circle without slackening speed. He had just about learned to accomplish this when Joaquin Felix came by and told him that Juan Valdés on San Emidio was ready to start the rodeo and for Salvador to go there. Salvador protested, saying he was getting along fine but in the end he went to San Emidio and Valdés sent him to the camp at Santiago. Salvador says that Antonio Felix was the top hand in that part of the country until Juan Gomez went there. One day Antonio roped a wild steer which got away with his reata. The only horse left in the corral that day was an enormous big footed animal used to pull a wagon though he had been ridden a few times. Juan came into camp and wanting to make himself useful, caught the big horse and saddled him, putting on a work bridle since the horse did not rein. He rode out and came upon the wild steer dragging Antonio's reata. He managed to get the big horse up to the steer and roped and hog tied it. Charlie Hudson, who spoke Spanish said, *"Que pasó con el famado Antonio Felix? Aqui llego su padre."* "What happened to the famed Antonio Felix? His master has arrived." And he remarked further "Half a man on a work-horse and he catches the wild steer." Juan has only one leg.

Antonio never forgave Juan and one day put him in charge of a bunch of cattle in a field where the floods had washed out the fences and some of the cattle were missing, but Juan saw that the fence was down and where the tracks of the missing cattle led. The next morning he arose at three and patched the fence. When it would hold stock he mounted his horse, followed the tracks and brought back the missing cattle. When Burt Snedden rode by he noticed the fences were mended and Juan did not get in bad after all.

Salvador says that the men on San Emidio were never paid for breaking a colt until George Brunk became foreman there, and then not until the vaquero had first roped a cow from the colt to prove he was a vaquero horse.

Prayers, Hogs and Snakes . . .

Vicente the vaquero says that if a turtle bites a person it will not release its hold until it hears a mule bray. He also contends that a mule will try seven times a day to throw its rider and will try every day at exactly the same time for seven days afterward. Vicente was one day riding a mule past a church. The padre was standing with a group of men at the foot of the steps. As Vicente passed, one of the men threw his hat under the mule which promptly kicked at it and lit into bucking. It bucked hard and long but the vaquero weathered the storm. When it had resumed a horizontal position again the padre said, "Son, if I had not prayed you would not have stayed on that mule," to which Vicente replied, "Padre, if I hadn't squeezed my legs hard your prayer would not have been heard."

Vicente was once camped in the mountains. For two nights a lion had prowled and screamed around his camp. A vaquero rode by and Vicente told him about the lion keeping him awake. The vaquero was alarmed and said, "You must leave this place because a lion always attacks on the third night. Leave your coat on a rock and tomorrow you will see."

The next morning the two men returned and found the coat torn to shreds.

In talking about characteristics of various animals, Vicente stated that hogs are the only domestic animals that snakes cannot harm; that hogs kill and eat all snakes and soon exterminate those in their vicinity. To kill a snake the hog kneels on it. If the snake strikes the knee it can do no harm as the knee is the only place on the hog's body where the poison cannot penetrate. He says further that a red steer has the best hide for making a reata and that if one stakes out the fresh hide when the sun is up and braids a six strand reata he will have a good one because a six strand is joined more easily than a four or eight strand.

Adolfo . . .

Adolfo says he was uncertain and afraid on his first day of work on Tejon Ranch. Don Luis Lopez took him under his wing and the boy rode away with the old gentleman to hunt wild cattle in the brushy Sierra. As they rode along, Don Luis said, "I am going to teach you to rope and tie wild cattle. You must learn if you are going to be a vaquero."

They had not ridden far when they came upon a big red steer which broke out of the brush and raced across a clearing. Don Luis gave chase and soon overtook him. He threw his loop and as it encircled the horns, he jerked it up snugly around their base. Adolfo then threw the slack around the steer's hind quarters, took a turn around the saddle horn and urged his horse on past the steer turning it heels over head. The steer's horns acted as a prop to hold the neck in a twist under the body and so prevented it from rising. The old vaquero dismounted and tied its legs together, the hoof of one leg against the hock of the other. Adolfo watched and expected to see the animal escape. A steer rises on its hind legs and it was easy enough for this one to get his hind quarters off the ground, but when it struggled to get its foot loose it would lose its balance and fall over. Don Luis said, "That steer will be there when we get back," and rode off to find another.

Adolfo's Mascadas . . .

Mascadas, scarfs, were made of bright colored silk and in early days dressy young vaqueros wore them tied loosely around the neck; that is, until one of the young men so adorned got into a fight and his antagonist caught hold of the scarf and while twisting it, choked him with one hand and beat him with the other, an experience which never failed to influence him to alter his style of dress.

In the days of his youth, Adolfo affected flaming red *mascadas* and this is how he came to eschew such "frippery" forever. One day the vaqueros came upon a number of cattle estrayed from a neighboring ranch. Since it was the practice to make such cattle pay for their keep by using them to improve their roping, the vaqueros took down their reatas and began "tying on" to the strays. They were having a very happy time when they heard cries for help. Adolfo, who had roped a steer some distance away, in taking his turns around the horn had caught the ends of his *mascada* under the rope and as the rope slipped the ends of the *mascada* wound like thread on a spool. He was gasping for breath when one of the men cut his good reata and saved him from choking to death. Ever afterward he abhorred this particular style of adornment and the sight of a scarf reminds him of the time he lost that good reata and almost lost his life as well.

Adolfo and the Beans . . .

One can see at a glance that Adolfo is an old vaquero. The hardships and broken bones suffered in a lifetime spent on cattle ranges have left their mark which time or circumstance can never erase. He is a member of that fraternity whose calling produces many thrills and dangers to life and limb; a class of men who have a propensity for excitement and welcome the caprices of their four-footed companions with an eager willingness for strife. This story may call for an exertion of faith, but it proves that Adolfo has absolute confidence in the vaquero's lucky star.

"Es cosa curiosa, it's a curious thing," he began, "how there is always a way to escape from the most serious predicament. I was working for Rolland Hill and had the Bear Valley camp when we ran out of beans. After a number of days without beans we began to crave them; no meal seemed complete without *frijolitos fritos.* My wife, in that mysterious way women on isolated ranches have of learning things, seemingly from the birds, said that Juan on Tejon Rancheria had harvested two sacks of beans and had one to sell.

"At that time I owned a sorrel horse that was mean, a silver overlaid *freno del pais,* spade bit with rawhide reins and silver mounted headstall and a full flower stamped Visalia saddle. I saddled the sorrel with this outfit early one morning and leading an old pack mare, I set out across Cummings Valley and down the mountain to Tejon Canyon. All the way down the trail the sorrel kept humping his back and looking for an excuse to buck. On arrival at the ranch I found Juan at home with a full gallon of wine. We had a drink and I made a bargain for the beans. Then we had another glass and another

was riding Mula. They had not gone far when
egan to buck. George could ride him, but the
ould buck, quit, and buck again. He kept this up
atarino, seeing the rider was worn out, told George
any one of his horses. Brunk refused because the
nted Mula ridden, but Catarino insisted and said
he boss kicked, he, being foreman, would explain.
ended up at Poso where, in his old age, he quit
. Years later when George quit riding he went to
r the State and was retired only a few years back.

des . . .

n the trail drive during the watches of the night
spectre of death, hovering over the tossing horns
eyed cattle, comes closest to the men who sleep
round near by. Cattle are often frightened and
wild stampede over the range, crushing every-
their path. The dread of being trampled to death
ost in the mind of the vaquero when he spreads
ets to bed down near a herd of cattle.

ory goes that in one season in the driest section
l California's range land, the vaqueros, riding
ing to spring, came upon large numbers of cattle
neras, or bands of horses, which bore the same
rand. The horses were all saddle marked and all
d. These animals overtaxed the capacities of the
hich supplied water. Cattle are placed on the
cording to the number or size of troughs and
f water the spring or well can supply. If there
any cattle none of them will get enough to drink.
press around the trough, licking the damp bot-
crowding will eventually break down the trough
hirsty animals tear down the fencing around

until the jug was empty. When I was ready to go I put the sack of beans on the pack mare without tying it, as I was too full of wine to do so, and started up the trail. The sorrel was rested and tried to buck, humping up when the rope touched his rump. I wasn't as careful as I had been on the way down; alcohol always makes one feel reckless. I tied the rope to the saddle horn and rode on. In going up the trail we came to a part that was divided by a huge rock. My horse went one way and the pack mare the other. I had to back my horse and in the process of this operation the rope got under the horse's tail. He started kicking and bucking and the pack mare pulled back and slid down over a bank. My horse was caught off balance and was pulled down on his side with his feet uphill against the rock. Luckily, I wasn't caught under him, but fell downhill and was knocked unconscious. When I regained my senses, I found the sorrel had somehow got up and was gone, but the pack mare was standing against the bank. I got up and picked up the lead rope and found it had been cut with a knife. From the tracks and marks on the ground I could see what had happened but could find no sign of a person and if some one had come to my rescue, why had he not stayed until I came to? I have thought many times of this incident but have never solved the mystery.

"The sack had fallen and the beans had burst out of the sack. I left them there, mounted the pack mare and rode toward home with many misgivings. I never expected to see the silver overlaid bit and rawhide reins again and I was sure the saddle would have turned and been kicked to pieces. My wife had a hunch that I was in trouble and had come to the Hill ranch in the buggy. She drove me home where I went to bed for six days. During that week a vaquero led the sorrel to my camp.

The Spanish bit, rawhide reins, and headstall were intact and the Visalia saddle didn't have a mark on it. By some miracle the horse had gotten up with the reins lying on his neck. He had gone to the ranch where the boys had caught him and removed the equipment. A few days later, I rode down to the rock where I had spilled the beans. It had rained during my spell in bed and the beans had sprouted and were ruined."

El Chileno . . .

During the early eighteen hundreds a fishing boat flying the United States flag put in at a tiny village on the coast of Chile and found an orphan boy. They took him aboard and brought him to the United States. It was either in Oregon or northern California that a German family took him in. When he was old enough to work, the boy drifted across the Sierra to Lake Tahoe and into Nevada, then following the eastern base of the Sierra Nevada, he drifted into Kern County and the southern San Joaquin. Here he worked for Jim Crocker until that worthy sold out to Miller and Lux. Although José Suniga spoke very good English, he was called "El Chileno" for he spoke Spanish in the Chilean idiom.

Catarino says that at that time all the ranches had the vicious custom of keeping a bad horse to try each new vaquero's riding ability. Catarino has forgotten the horse's name but remembers that he was *retino,* brown, and very fast. He says he doesn't know where the Chileno learned to use the *barbiquejo,* an Indian system of bridling, though the word *barbiquejo,* like *jaquima,* hackamore, comes to us from the Moor. As the Chileno had roamed about extensively, he could have picked up the knowledge from the Shoshones or Paiutes of Nevada or from the Oregon Indians.

On the Chileno's first day on
the brown horse. When the hor
blinds on. Then, as the men v
on the *barbiquejo,* a thong ti
to which reins are attached.

He mounted, and the horse
given him a bad time walked

In those days there were n
were several *quadrileros,* a
meaning hip. These men wer
They would dismount, rope a
their hip, hold him until he ch
the vaqueros surrounded the
mounted with their reatas.
his reata go and a colt broke
been told that the horse he
he put the brown after the
When in reach of the draggi
over his horse's right eye to k
then he leaned down and ca
turns around the horn and
tween his leg and the saddle
began to buck and with his
horse spun first to the righ
putting the rope over the h
in the direction of the runaw

After finishing the story a
went on to say that there
liked very well: Ernest, Ha
worked with him years ag
fine looking bay horse at S
he had abnormally long e
Catarino in charge, start

Georg
Mula
horse
until C
to take
boss w
that if

Mul
bucking
work fo

Stampe

It is
that th
of wild
on the
rush in
thing in
is upper
his blan

The s
of centr
from spr
and *cap*
strange
were sho
springs
ranges a
amount
are too m
They wil
tom. Thi
and the

the spring, making a mud hole of it and stopping the flow of water.

The cattlemen immediately sent their riders to find the men in charge of these stray animals. The vaqueros searched in all directions but for all their hard riding, the drivers of this strange herd or the riders of the strange horses could not be found. The ranchers hurriedly gathered the strays, which made a large herd, and started to drive them to a river where the water supply was sufficient for the needs of these unwelcome guests. They were well started when a number of men arrived. These newcomers were in search of these cattle and the drivers because the outfit had not arrived at its destination. After being acquainted with the strange circumstance, they began a search back over the route the herd had come. After some days, two men riding along the foot of an overhanging bluff, came upon the scattered remains of a camp. Among the broken saddles, crushed pots and pans and torn bedding were the bodies of a number of men. The condition of the bodies told plainly that a multitude of hoofs had trampled the life out of the men who had gone to sleep under the overhanging bluff. Farther on, four horses and their riders had met death in the same way, having been overtaken by the stampeding cattle. After the bodies had been given Christian burial, the owners moved the herd on to their home range.

All vaqueros have, at one time or another, been badly frightened when they have been awakened from a sound sleep with the cry, *El ganado, el ganado,"* (the cattle, the cattle) which sent them running in the dark, in their bare feet, for their lives.

José Juan had come awake from a toil deadened sleep at the cry of warning. As he sat up, the knee of one of his companions hit him on the nose and broke it. As he gained his feet, a horse running in panic, struck him, hurting his hip. Unable to walk, he crawled to a tree where the other men had crowded behind its meager protection, pushing each other away from its slender trunk. José Juan said it was the bridle that saved him. The roar of the hoofs was loud and close when he reached the tree and tried to rise up enough to grasp a branch. In the dark his groping hand encountered a bridle which hung from the branch he was trying to reach. He does not know how he got the strength to pull himself into the tree, but he was sitting on the limb when the cattle swept by, missing the tree and the men crouched behind it. The men's equipment was a total loss; the bridle that José Juan had used to save his life was the only whole thing left. There followed many weary days of riding bareback, after the tough job of catching the scattered horses, and gathering the cattle, which were scattered to the four winds, before even a part of the gear needed could be acquired again.

Though most *stampidas* ended in tragedy, Raoul Martinez tells of a boy who would try to snatch a little more sleep when his turn came to go on night herd. The men who suffered from his laziness tied three pack saddles together, dragged them over his body, yelling that the cattle were stampeding. The boy jumped up with his canvas over his head and ran into a cactus. He was cured of his laziness.

The Vaquero's Sense of Humor . . .

One of the characteristics of the vaquero, developed no doubt from his reaction to his daily fare of risk and danger, is a definite sense of humor. He never failed to

Guadalupe "Lupe" Gomez

see the funny side of many of the day's petty accidents; nor did the vaqueros ever withhold their laughter when some diverting incident occurred to excite it regardless of how much vexation it caused the unhappy victim.

Lupe Gomez, a hale and hearty old man who has laughed often and long, is a good story teller and all of his stories have an element of humor in their makeup.

Comical things happened every day and it is a curious circumstance, but the memories which are treasured by the old vaqueros are not of the times when they suffered hardship but of the times when they laughed. When they tell of the old days it is the humorous situations with horses which are recalled.

Salvador Carmelo and Lupe Gomez were going back in retrospect to their days on Tejon when gathering cattle meant leading wild steers out of the Sierra and yoking them to tame oxen. Lupe asked, "Do you remember the sorrel horse El Washington? He was very gentle. Juan, who had only one leg, had ridden him for years when one morning I borrowed him. As I mounted, he started bucking, the first time in his life he had ever done so. He wasn't used to having a leg thrown over him."

Salvador answered, "Yes, and do you recall Ralph Bowers? He was a good vaquero and never rode any but a good horse. One day he was preparing to doctor a cow and was whirling a big loop when the horse he was riding switched his tail and caught the *reata* under it. The horse clamped his tail down on the rope, holding it as in a vise. The sudden jerk caused Bowers to drive his thumb into his own eye with a good deal of force. With his eye streaming, he dismounted, took his knife out of his pocket, opened it and groping for the offending tail caught it and cut it off as close to the dock as he could. The horse's ludicrous appearance and Bowers' show of temper brought roars of laughter from the vaqueros."

Compadres . . .

It is when the ride is well begun and the horses are settled to a running walk, keeping pace and swinging their heads, that the *cuentos* begin, if one is lucky enough to ride with a story teller like Adolfo. The stories of necessity must be short for rides are often interrupted by some unforseen incident.

Our horses were walking steadily as we passed the last of the *casas* and Adolfo pointed and said, "One of my godsons lives in that house." I had often heard him mention this or that *compadre* and I remarked that he must have a lot of godsons. He explained, "A man must have at least six godsons before he can enter the pearly gates. *Compadres* are the only ones who meet in the next world, so it behooves a man to have many godsons so that he may have many friends in the hereafter. That is what people say, but only God knows for sure what becomes of us."

When a person asks some one to sponsor his child, it means he has been chosen for a brother in this world and the next. A man's last recourse is his compadre, for compadres never go back on each other or refuse a request from one to another, for God pardons everything but a sin against a *compadre.*

A child's godfather must be ready to accept his godson as his own in the event the child should lose his parents. Perhaps this custom started when Moor and Christian fought for the possession of Spain; a parent could die a little easier if he knew his child would be cared for.

In the old days it was said that the owners of grants

of land were all compadres and that governors selected the best land for their children's godfathers.

Compadres never use the familiar *tu* but address each other formally as *usted.*

The *paisanos,* or countrymen, nowadays are often too *agringado,* Yankeefied, to follow in the ways of their ancestors, but in the old days even the gringos observed the customs of the *hijos del pais* and quite often sponsored one of the numerous *niños* and became a *compadre* to some *paisano.*

El Viejito . . .

In the old days, *muchachos malcriados,* ill bred youngsters, were not often met with, if at all. One of the reasons why people respected their elders might have been the result of the many stories where this particular virtue was the main theme. There were few books in California and fewer people who could read them. In those days the *cuento* was relied upon to teach manners and this story is one that has been told many times through the years.

The ranch was known far and wide for its mean horses. When the ragged old man with his wizened little wife went there and asked for work, the *mayordomo* pitied them. Out of respect for the *viejito's* white hair, he assigned him a *jacalito,* little hut, and gave him a spavined, ring-boned old mare to ride when herding the goats. The old one accepted this work and the decrepit mount humbly and thankfully and took such good care of her that it was not long before her ribs filled out and her coat gleamed in the sunlight. In the morning the men would all grin when the *viejito* rode out on the old mare. She would mince out of the corral with a hump in her back; but only for a few yards, when old age and infirmities would bring her back to a more sober behavior.

The time for working the bands of horses comes in winter, long after the rodeos are over and the cattle have drifted out of the sierra into the warm lowlands. The saddle stock is culled; the older horses turned out to rest through the winter, while the four year olds are brought in and issued to the men who are to break them.

One cold morning the corral was jammed with wild-eyed broncos and the men were gathered in a group near the gate to wait for the colts to be roped and allotted to them. As each animal was caught the *caporal* called a vaquero who led it away and tied it up. The *viejito* had joined the vaqueros when the work began and stood patiently by until each man had been issued four broncos each and there was but one horse left in the corral, a *bayo lobo,* or buckskin with a coating of light hair over his black mane and tail. He had been ridden and spoiled the year before and had been turned out; sometimes horses forget about bucking after running out a year or two. The *mayordomo* was leaving when the *viejito,* seeing he was to be left out of the picture, approached him and said, "Señor, I have not been issued a *potro,* (colt)." The mayordomo was of the old school who show respect to their elders so he spoke gently.

"*Tio,* uncle," he said, "the nature of your work does not permit the riding of *caballos broncos* and for that reason I did not call on you to take one."

The *viejito* with *mucha tristesa,* very sadly, said, "I would like to earn the extra money riding the *potro* will bring."

The *mayordomo* pondered. It did not look well to give the *viejito* the spoiled horse he had kept from younger men; on the other hand to refuse him a colt would be as much as to say the old man was worthless, so much against his will the *mayordomo* told the *viejito* to take the buckskin.

The *viejito* immediately removed the saddle from the old mare and carried it into the corral where he untied the *tapa ojos* (blinds) and *jaquima* he had brought. The *caporal* threw a loop around the bronco's neck and led him into the empty corral, where, after he had choked down and fallen, the old man fastened the *tapa ojos* and *jaquima* on his head. When the loop was removed the colt stood up and the *viejito* put the saddle on, cinching it up tightly.

By that time, the men, sensing excitement, had all returned to the corral and were talking among themselves. One of them, seeing what the *viejito* was up to went up to him as he was reaching for the stirrup and said, *"Mira,* Don Luisito, here are four reales. Each of us will give you this amount so you need not ride the horse."

The old man made no answer until he had mounted the trembling buckskin and got a firm grip on the reins, then he said, "I will show you what to do with your money. Put a coin on each stirrup."

The vaquero put his coin on one stirrup while another man put a coin on the other. Then the old man put a foot on each coin and raised the blinds. The horse bucked long and hard and when he quit and stood panting the old man raised his feet and there were the coins where they had been placed. Thus did the little old man gain the respect of the hard bitten vaqueros on that ranch.

Curanderos . . .

Back in the days before the gringo had brought doctors, drugstores and hospitals to the San Joaquin, the *curandero,* or healer was depended upon to cure the sick.

Though they were often *matasanos,* (literally, killers of the healthy) some of them were quite skillful. A curandero was anyone who professed a knowledge of healing, even if the extent of his practice was limited to *flotadas.* A *flotada,* or massage, was given for every pain or ache from pneumonia to simple earache.

Every kind of grease from beef tallow to rattlesnake oil was used as an ointment. The latter was secured in the following manner: first, a snake was found asleep stretched at full length and before it could coil, get angry and bite, it was struck on the head and killed. Next, six inches of the body was cut from both ends because the poison is in the head and tail. The snake was then cleaned, the grease removed and put in a pan or dish and set on the roof of the house for the sun to render, as the oil would be useless if rendered by artificial heat. The remainder, which consisted of the vertebrae, was hung up to dry. This was good for people who had sores; a small portion of the jerked snake would be pounded to a powder and sprinkled over the sick person's food.

The snake oil was best for deafness. The warmed oil was poured in the ears and the ears plugged. The oil would penetrate the inner parts of the ear and cure this distressing condition.

Salt was considered a cureall and oak ashes were taken internally as an anti-acid.

Cobwebs were applied to a wound to stop the flow of blood, and sugar mixed with pounded charcoal and put in the cut would heal it.

Cactus, besides being used for food was toasted and used as a poultice.

The old reliable among the herbs was *yerba del pasmo,* which, when boiled was used to wash wounds and could be found in the store house of all ranches. Chili also had medicinal value. Whenever a paisano had an *empache* (indigestion) the curandero removed the seeds from dried red chilis until he had a spoonful. These were washed in nine waters then given to the patient to swallow. The seeds would attach themselves to the *empache* and remove it from the spot to which it was glued. Vaqueros believed that an *empache* was an undigested morsel of food which stuck to the walls of the stomach and caused the pains many were bothered with. *Yerba del pasmo* is known botanically as Waltheria Americana.

A sure cure for acute indigestion, though somewhat drastic, was the rawhide riata. There was a young vaquero who owed his life to such a rope and the stern measures of a *mayordomo*. Though this youngster never cared to have his experience mentioned in after years, this is how it came about:

The vaquero crew had left early one morning with a herd of cattle to make a *jornado,* or drive, across a long stretch of desert to a canyon where there was a ranch and some irrigated fields. It was the evening of a long summer day when they arrived at the ranch to find the *jacal* (hut) deserted. The door was open but nothing to eat could be found. The hut had not been empty long, however, because the *milpa* (cornfield) was green and had been recently irrigated.

After the cattle had been watered and had drifted up the canyon to feed, the vaqueros gathered ears of corn. They built several fires and ate the *elotes,* corn, until their hunger had been dulled. One of their number, despite warnings from the older men, ate his fill of the corn, to his sorrow.

Vaqueros go to sleep early but they had not slept long

Dan Castro

before they were awakened by the boy who was crying in pain. The mayordomo studied the situation briefly and decided on a course of action. He ordered one of the men to bring a riata. He took the rope and threw it over a low hanging limb of a giant fig tree. He then pulled both ends together, and ordered the men to bring the boy. When they were under the branch he told them to stand the boy on his head and hold his feet upright. The mayordomo then tied both ends of the riata to the boy's ankles. He took a stout stick, put it between the ropes and turned the boy round and round until the riata was was twisted enough to raise him off the ground.

Then the mayordomo suddenly released the stick and the twisted rope spun so rapidly it made the boy dizzy. This proved to be an effective emetic, and the boy, having gotten rid of the offending corn, was relieved of his pain.

Dan Castro . . .

Dan Castro is seventy-four and has said goodby to the vaquero's calling after sixty years of riding the range and now some younger man can use his rawhide reata, Spanish bit and Visalia saddle.

He began riding for Miller and Lux on the Maple Ranch in the nineties and ended up in 1956 on the Jack Ranch at Cholame.

One of the good vaqueros of the past era, Dan has observed and studied the animals he has handled and as a consequence knows many of their characteristics. My discovery of Dan's talent as a story teller came about because Tilton Tracy bought a buffalo. I had been out to Tracy's and I told Dan how fierce the buffalo was and how unlike a cow for he will not be driven nor will he run away from a dog, but will turn to face his enemy and hold his ground, butting, striking and kicking as through the centuries buffalos have fought off hungry wolves; and though born in some fenced in area of our Southwest, this atom-age anachronism is still a wilding. He will not mix with the ponderous beef steers in the feed lots and butts them away from the manger. Perhaps he despises them for their tameness, animals fit only for the slaughter.

Dan then told me of the ostriches the Tracys had owned and the last time he had gone to a rodeo there which had been during World War One. It was a benefit show and Dan won eighty-one dollars riding bucking broncos. He turned the money over to the Red Cross for Dan could not serve in the army because of a bad leg and the Castros are good citizens.

In telling his stories Dan raises his voice for emphasis and repeats, like this: "a good horse, a good horse," the last good said at the top of his voice would mean a most outstanding horse.

One of the unforgettable experiences of his lifetime was the roundup of the Tule elk in the nineties. Three carloads were shipped out of Buttonwillow Ranch from the corrals at Lokern and on the morning of the elk rodeo, three hundred mounted men showed up to make the drive. Dan is careful to make a distinction; some were vaqueros and some were dudes. Miller and Lux and the Kern County Land Company had sent all their available men.

On the west side of the valley was a slough called the Dredger which was too wide and deep to cross without swimming. The dudes were lined up along the east side of the Dredger, while the vaqueroes began the drive at the southern extremity of the Elk Hills; a wing extended from the corrals at Lokern for a distance of three miles.

Elk, Dan says, always run in a straight line and will not swerve from anything in their path, but will go over it or through it. This characteristic made them easy to rope as a man would know where they would pass and could waylay and rope one for there were very few if any horses fleet enough to overtake one by racing after it. In fact, it took "a good horse, a GOOD horse" to even stay in their dust.

The Miller horses, which at that time were S Wrench Oregon mustangs, all weighing less than a thousand pounds, proved to be able to follow the elk better than any of the horses from the other ranches.

A lot of dudes met with disaster that day. The Dredger had to be crossed and many of the horses had never been in deep water in their lives and would try to cross by walking on the bottom of the slough, but would soon come up choking and gasping and start swimming in a most clumsy manner. Of course the riders would have to leave them and strike out for themselves. It is not true that horses know by instinct what to do in water. Dan says they adapt themselves and learn as humans do.

Droves of elk would hit the water so hard and fast that they would push it to the side and make a path through it. Some of the smarter dudes learned to stay close behind and cross before the water flowed back and walk between two walls of water as the Israelites did when they crossed the Red Sea.

It was not hard to hold an elk once it was roped. It would give one or two hard jerks, then quit. A vaquero could then give his rope to a dude, take the dude's rope and catch another elk. In this way some use could be made of the dudes. Dan says that one day some government men were going to take elk away and Henry Miller stopped them by saying, "Those elk don't belong to the government. They belong to me and are on my land." The men went away and never came back.

Jim Ogden, the ranch superintendent, would pay ten dollars for each elk's tooth, and since the vaqueros of that day were wizards with the reata there were few bull elk, which, once encountered, were left to go their way with their "bridle teeth" left in their mouths. Each vaquero carried a piece of sack twine. When the elk was down he would take a couple of half hitches around the tooth and jerk it out. The teeth, Dan says, didn't seem to be too firmly set.

One day Dan was riding with Rafael Cuen. They were forcing their horses through a thick growth of sunflowers when an elk started up. Dan, despite the yells of Rafael to "let him go," took after him. He had his rope ready to throw when the elk became entangled in the sunflowers and fell. Seeing him securely hobbled, Dan jumped off his horse and running to the elk took the sack twine, jerked out a tooth and mounted his horse again. I asked, "What if he had got up?" "He would have killed me," he answered, "they were dangerous."

Speaking of his first job with Miller and Lux, Dan said he was riding fence one day when Miller drove up beside him and asked how many cattle were in that field. Dan said he did not know. "What," Miller roared, "You don't know! You should always know how many cattle you have." That evening Miller raised cain at headquarters. After that Dan was always told how many head of cattle he should have.

I asked Dan if Henry Miller spoke Spanish and he said, "You bet he could and a lot better than I can."

Dan said that when he got tired of riding after cattle

he would get a transfer to the hog department and become a hog buckeroo. The hogs were handled on horseback in the same way as cattle. Each vaquero, or let us say *cochero (coche* means hog in the dialect spoken in California) had a string of horses trained to work hogs, the horses being smaller than cow-horses. There were thousands of hogs which had reverted to the wild state ranging in the Tule marshes and sloughs on the west side of the southern San Joaquin, and all were fierce fighters. The horses were trained to hunt them and became very expert in following them by the sound of their progress through the tules. One man alone could not very well handle a wild hog. It was a job for two; after a hog was roped one man could dismount and tie it while the other held it.

Boars could be led out into the open because their tusks curved back and would hold a loop from slipping when it was put around their upper jaw.

A wagon with a rack built on to hold the hogs followed the crew and when a hog was caught it was placed in this wagon. When the rack was full of squealing hogs they were hauled to a feed pen to be fattened for market. If one had the bad luck to be caught afoot by a wild hog, the only way to escape death or injury was to mount and ride for it. Through some freakish whim of nature a good many of the hogs had long hair and by getting a good hair hold, a man could ride one. After a few jumps it would get out of breath, choke and fall down, then the man on its back could tie it up. "But he had better ride it," Dan said, "because if he fell off it would be too bad. The hog would have him down." In this way, day after day and month after month, the hog crew hunted and roped hogs and brought them into the feed pen.

Forty years later, in 1936, Dan was riding for Roland Hill and along the San Joaquin River near Merced he found Miller and Lux hogs that had survived in the tules. They were so inbred that some of them had a solid hoof, that is, their hoofs were not cloven like normal hogs. In the time he spent working for Hill, Dan caught the last of them. One little pig was so fierce that to keep him from biting, Dan rolled him up in a saddle bag, wrapped and tied a string around him and laid the bundle across the saddle and rode home with it.

One time, an old boar with long curved tusks was caught after a long hard chase. Dan put him in a truck and when he got to the slaughter house, he wouldn't get out so Dan tied the hog's lead rope to a corral post and drove out from under him. The manager of the abbatoir gave Dan a pound of dressed meat for each pound of live hog. Why, Dan does not know.

Years ago, Dan's family ran a band of brood mares on Breckenridge Mountain. Though it was a large band, the colt crop would be only four or five a year. They were being killed by the numerous mountain lions. One day a man from the mountains of Sonora came along and told them that they would not lose the colts if they turned mules in with the mares. So one of the Castros went south of the border and bought some twenty mule colts which were turned out with the mares. From that day on they lost no more colts. The mules must have fought the lions away.

For years afterward anyone who put on a rodeo could borrow the Castro mules. Years of being used as bucking stock so improved the little fellows' bucking ability that the last time they were loaned to a rodeo the cowboys could do nothing with them.

Dan remembers one long last ride he made. Rafael Cuen, the cattle boss, had given his vaqueros a few days off to celebrate the Fourth of July. The entire crew had ridden into Bakersfield and were enjoying themselves when Cuen rushed into town and ordered them to get their horses and ride to the Bell ranch. A levee had broken and the cattle had to be moved out. The vaqueros hurried out of town and galloped all the way, some eighteen miles, and rounded up the cattle. By the time the last steer was driven out of the low land the water was so high that in some places the horses were swimming over the barbed wire fences, but they didn't lose a cow.

Vaqueros . . .

Fiction writers place the cowboys they depict in settings they have never seen, form odd mental pictures and write lies and nonsense about them. They credit them with qualities they do not possess and attribute habits to them of which they have no knowledge. They are, however, in their true environment, colorful and picturesque figures; daredevil spirits in love with their craft and ready for anything that promises adventure. Often ignorant of books, they are educated in experience, with a touch of the Arab *magnifico* and of Don Quijote, but with a stout strain of the heroic in their make-up; characteristics which have made them distinct from men in all other walks of life.

Vaqueros are a race of men who have disappeared and though venerable grave-faced men have told interesting accounts of their adventurous lives, their accounts were as far removed from the imaginary tales of fiction writers as day is from night.

Writers have committed many offenses against the men who herded cattle for a living, but the worst crime of all is the horrible language they put in their characters' mouths which is the least typical of all.

A Rough Ride . . .

José, the vaquero, had good reason to be bowlegged. He began to ride before he could walk, rode a bucking mule and fell off a horse while still an infant in arms. And this is how it came about.

On the day José was to be baptised his parents started celebrating early in the morning and by the time they were ready to go to church, several bottles had been emptied. They put the baby in a *vatea* and mounted their horses. The mother carried José. A *vatea* is a trough-shaped vessel used to hold the cornmeal from which tortillas are made. They had ridden some distance when the father suddenly spurred his horse and dashed forward. The mother raced after him and while trying to overtake him, dropped José out of his makeshift cradle, unaware that she had lost him.

The infant fell on a bush and was not hurt. He had lain there for some time when a neighbor riding a bronco mule came down the trail. When the rider came opposite the bush the baby let out a wail and the mule began to buck. When the storm had subsided, the man dismounted, drew his pistol and cautiously approaching the thicket, found José. He blindfolded the mule, took the baby in his arms and mounted. The animal made a few jumps but the vaquero and the baby stayed aboard. They rode on and met the parents who had missed the baby on arriving at the church.

Willie Tevis . . .

We were on our way to Bakersfield, driving toward Palm Springs. We passed through groves of date palms and around a curve came upon the polo player and stopped, as we have through our lifetime stopped when we have seen a good horse, a good rider, a good saddle or bridle bit, or a good picture such as a Russell or a Remington.

He was making a few practice strokes and the way he handled horse and mallet spoke of many years of experience and brought back memories of an Argentine team riding *hell for leather* over a field against the best of America's players. He was quite elderly, but his seat on the grey horse and the way he brought him into position for the stroke of the mallet told of lessons learned from experts in Spanish equitation. I had known him in 1918 or thereabouts and since heard many tales of his prowess but had not seen him again so didn't recognize him. In a few minutes we renewed our acquaintance and I found he was the man the foreman on the Land Company ranch saved the bad horses for; he used them for practice riding. He was the man who rode a bucking horse, using an English saddle at Madison Square Garden and again at Salinas. He was the man who rode lone against the cavalry in the longest Pony Express ride in history and outrode them all. Mary Pickford made the presentation of a gold medal for this feat. He still carries it on his wrist.

He rode against the finest polo teams ever to come to the United States and played with skill and sportsmanlike courage. The polo player was Willie Tevis, gentleman rider, cavalry officer and one of Kern County's greatest horsemen. It was a coincidence, but when he dismounted there was a man waiting with outstretched hand, the light of hero worship in his eyes and a smile of pride in knowing him. "Mr. Tevis," he said, "I am the man who held the horses for you when you rode the Pony Express race. I am Don James." Mr. Tevis remembered him and said, "Yes, your brother Earl helped me saddle the horse I rode at San Emidio."

"It brings to mind," Mr. Tevis went on, "a Kern County Land Company rider who would throw a horse, saddle him, mount while he was on the ground and ride him to a standstill. I was a little boy then and my one ambition was to ride like this vaquero. Many years later I went to stay at the old house at San Emidio. Juan Valdez, the foreman, told me he had an outlaw horse he wanted me to ride. The horse was brought out and we saddled him. The man holding him by the ears was the bronco buster I had vowed to emulate. I remember he offered me the use of his quirt. He was crippled and unable to ride any more and there I was, riding as he had once ridden."

I thought and thought; man hurt by a horse, good vaquero, couldn't ride any more, and suddenly it struck. It could be only one man. I asked Mr. Tevis, "Wasn't that vaquero's name Turner Rose?" and he answered, "Yes, yes, that's it, Turner Rose."

Mr. Tevis asked me to go with him and watch the game he was to play in that afternoon. It was a pleasure to watch this great-grandson of a king's soldier and grandson of the last man of Spanish blood to govern California play in that inimitable style of his. Whenever the ball flew toward the opposing goal post, the veteran would be there waiting for it as though some instinct told him which way it would come. After the game was over, his daughter asked him to ride her horse. She wanted us to see the animal spin and back. She told us proudly, and justly so, that this horse, a black, was the last horse trained by Larry Lavers, one of Kern County's finest reinsmen and a son of pioneers.

Mr. Tevis told me of the time Buffalo Bill's Wild West

Show came to Bakersfield and Buford Fox, his best bronco rider, came out to Stockdale to give the vaqueros a lesson in riding bucking horses. The vaqueros brought out and saddled Siwash, the outlaw. When it was all over it was Siwash who had given the lesson and the best bronco rider from the Wild West Show was in the hospital. Siwash remained unridden until Willie Tevis took a sitting and rode him to a standstill.

Diablura . . .

Once upon a time there was a man who had become rich through the practice of black magic. On top of a steep hill there was a mesa where he practised his arts. One day he told his favorite *caporal* to get ready as they were soon going to a Fiesta of Bulls. When the day came, they journeyed to the foot of the mesa where the magician told his *caporal* to close his eyes and not to open them until he was told to. When they had climbed to the top of the hill, the *caporal* was told to open his eyes. He found himself looking into an arena where the bulls had only one eye set in their foreheads and one horn projecting above the eye. The *caporal* said, "I see cattle with only one eye," whereupon his boss said, "There are stranger things than this to see." The *caporal* said, "I feel a great heat and I am afraid. His boss replied, "You will remain here as a punishment for stealing so many cattle." Thus the *caporal* was doomed to stay and dodge bulls for having taken that which did not belong to him.

The Piedra Pintada . . .

Far to the west of Bakersfield, across the valley and over the Temblor Range, lie the Carriso Plains. Overlooking this plateau and standing on an eminence in solitary grandeur is the Painted Rock (Piedra Pintada). It is called so because at some distant age the artists of a long gone race, migrating from the frozen north, left their indelible paintings in the shallow caves of the rock.

The Piedra Pintada seems to be the crater of a long extinct volcano which has been covered with earth until only the top remains above the level of the plain, making a hollow circle. There is a narrow opening into the crater.

The early Spaniards found it very useful as a corral when they ran the wild mustangs that abounded on the plains. It was a hiding place for Murrieta and in later years for Vasquez and Chavez, so legend says.

A short distance away is a spring of bad but usable water in an otherwise thirsty land. Because of this vaqueros have used the rock as a camping place since it was first discovered.

Over the years there have persisted stories of strange lights seen there by the campers who told of having heard the rumbling of phantom bands of horses and the crack of the *ramal* as goblin vaqueros struck their chap legs to urge on their charges.

There is a story told of a young cattleman named Jesús who rode the ranges constantly on the look for *lepes,* motherless calves, which he would catch and put on the few cows he kept for that purpose. They usually had two calves on one cow and often four or five *lepes* were suckled by one cow. In this way he gathered a herd slowly but surely.

One day there came into Bakersfield a gringo who spoke fluent Spanish. A close observer would have noticed that this man had learned the tongue in Sonora for his mannerisms of speech were those common in that part of Mexico. This gringo was even a shade more inquisitive than the general run of them. He asked many questions and favored the old vaqueros who were *muy platicadores,* very talkative, with his company and learned of the vaquero Jesús who knew the Carriso Plains so thoroughly and had camped at the rock many times. The gringo hired Jesús to go with him to Painted Rock. They set out in a buckboard and on the second day made camp in the old crater. The next morning the gringo said, "Jesús, find me a spot in the opening there where there are rocks of the same color," which Jesús did. Then the gringo said, "Now find some rocks which resemble sheep. They should be north of the opening into the crater." Jesús, knowing where to look, pointed them out. The gringo then paced out a distance from each point and said, "Now find the figure of a serpent in the paintings." Jesús went and stood by the figure. Then measuring his angles from rocks to figure, the gringo pointed to a crevice and said, "Jesús, there is a crucifix in that crevice. Get it." The vaquero found the article under a well-fitted slab of rock and gave it to his employer who said, "You may think this is gold, but it is not." He twisted the crucifix apart and revealed a paper folded to fit into the shaft of the cross.

The paper was a yellowed chart or map which broke into pieces when he tried to unfold it. The man had come prepared, and all through the day he sat and pasted the pieces of the map on a large cardboard and traced out the faded words and drawings. It was late afternoon when, with a satisfied look on his face he put the map away and helped Jesús prepare supper. But that was as far as the gringo was to go. Perhaps it was the nervous tension or the vile water which made him ill. That night the boy was awakened by his employer moaning in pain. Seeing he was violently ill, Jesús broke camp and in a forced drive took him back to Bakersfield and a hospital. He waited in Bakersfield but the gringo never contacted him again and he went back to riding the range. After a time he began buying large numbers of cattle and lo, he was a rich man. But there seems to be a curse on buried treasure and a rule that he who finds it never can keep it. The gringo, after all his labor of searching out the people in Sonora who knew of the treasure and his efforts here, never profited for his pains. And Jesús, who found the gold, in time lost all his cattle and ended his life as he had begun it, a poor man.

Crafty Indians . . .

This little tale of Indian cunning and enterprise comes from the time of the viceroys. In those days the number of people who could read or write were so few among both Spaniards and Indians it was the custom for tradesmen to use objects to advertise their wears instead of the written word. The shoesmaker hung out a boot, the blacksmith displayed a horseshoe and so on.

The man who had the monopoly on saddle making, Don Alonzo Martínez, hung a saddle tree over his shop door. One day Don Alonzo left his shop unattended to go to dinner and when he returned the saddle tree was missing. Some Indians who had been wating for the opportunity had stolen it.

A few days later while the saddle maker was again absent, the tree appeared over the door. He had almost forgotten the incident when some eight or ten days later an Indian appeared with a number of trees which were exact imitations of those made by the Spaniard and offered them for sale. It was then that Don Alfonso's fury knew no bounds for the art of saddle making was no longer the prerogative of the Spaniards.

Don Luis Lopez . . .

Catarino Reese had been talking about old-timers, so I asked him about Don Luis Lopez who had been an *angeleno* and so my *paisano*. He said there were men who had excelled Don Luis in some phases of the vaquero work, but as a reinsman he had no peer. It was in his day that the Indian, never before allowed to ride a horse, was to assume the prerogative of the *gente de razon* and adopt the bit and spurs of the Spanish cavalryman and ride the range.

Once, Catarino met the old gentleman in town and Don Luis invited him to see a horse he had at his place somewhere near the old Tracy home. The bay horse was woolly and poor and while Catarino didn't care much for his looks, he praised him a little out of politeness.

The seasons changed and one day Catarino was walking toward town when he heard a flurry of hoof beats and Don Luis was upon him. He would spin the horse to the right, then to the left, then circle around Catarino and when his back was exposed, Don Luis would strike him with his *ramal*. Catarino was brought up in an era when children were *bien creados*, well bred, so, despite the whacks he had received, he thanked Don Luis for showing the horse. The old man took a liking to the boy and one day when they were riding together, Don Luis took a sudden dislike to Catarino's horse and told him to turn the *mugre*, refuse, out and take one of his horses. When a man offers one a good horse, he offers his heart and such a gift from Don Luis put the boy in the seventh heaven of delight. He caught the horse and addressing the old man as *tio,* uncle, out of respect for his skill and age, he asked him to tie the *falsa rienda,* rope rein, to the *jaquima,* because Don Luis knew how the horse had been ridden. Taking lessons from such men accounts for the fact that Catarino's horses were of the best the Kern County Land Company owned.

One day back in town Catarino saw Don Luis sitting on the bay, fat and sleek now, in front of a saloon on 19th Street. A big burly man came out of the saloon. He had been drinking heavily and he walked up to the horse and made a grab for the reins. The old man backed the horse out of range and said, "I don't like any one to take hold of my horse," but the drunk made another lunge. Don Luis spun the horse away and seeing the man meant trouble, touched the horse with the spur and the animal sprang forward, knocking the man down with the point of his shoulder. The man rose only to be knocked down again and again until Bert Tibbets came out of the saloon and asked, *"Compadre, que pasa con Usted?"* The man, sober now, took this opportunity to crawl away through the saloon.

This blow with the horse is called *testerazo* and is one of the main requirements of a good horse. With a horse one could down a steer or bull when the need arose to save a companion in danger.

Don Francisco . . .

Don Francisco was a small boy when he began to worship the swaggering riders on the ranch where he was born and to envy them their bowed legs. He decided very early to ride horses until his legs should be shaped like those of the oldest vaquero.

His first job on the ranch was to herd oxen, and since

each ox had to be lassoed, he learned to throw a loop. He became expert because necessity is the mother of industry and because the rope was his only toy. Since he never went to school, he played with the rope through his childhood.

He learned to ride and to refrain from using bad language very young and it came about in this way. The mares and colts on the ranch were kept in a large pasture. To get water they passed through a lane which went by the house and down to the river. Trees lined this lane and it was here the horses spent the hot part of the day with their heads at each other's rumps and their tails switching flies.

The lane had a couple of gates which when closed made a corral. Don Francisco and his brother would climb a tree and sit on a limb. When a mare passed underneath, one of the boys would drop down on the animal's back, get hold of the mane and try to stay on. Sometimes they rode their mounts, sometimes they were thrown. But whatever the immediate results, they in time became riders.

One moonlight night they were at their practice, when the mare Don Francisco was riding bucked harder than usual and the boy yelled out, saying what vaqueros say when a horse bucks. When the fun was over the boys sneaked into the house to find their father waiting for them. "So," the sire said, "you have learned cursing." He led Don Francisco and his brother outside to a stout post, put the boys facing each other with the post between them and tied their wrists together. He then doubled his riata and brought it down across a boy's back. The boy jumped and in doing so, brought his brother into position for the next blow who in turn jumped and put the other boy in position again. So without too much effort the boys' father thrashed them most thoroughly. Don Francisco says that he was always careful about his language afterward.

Mark Parks . . .

I met Mark Parks for the first time some forty years ago and each of our meetings over this period of time have been impressed on my memory for Mark has something in his character that compels one to remember him.

It was a cold wet day in the fall of 1917 or '18 and we were driving some cows and calves along the highway from Buttonwillow toward Temblor. The cattle moved slowly, and being kids, we had our ropes down and were making practice throws at the calves' hind legs when the bosses, Mark Parks and Rafael Cuen, drove up in a buckboard. They had been watching us rope the calves and Parks said crossly, "Get the cattle out of the way, kid, so we can get by," and urged the team on past the cattle.

Some twenty years later in the middle thirties we were loading cattle in a field on the McKittrick Ranch. As there was no chute, we held the cattle in a corner; we would rope one, lead it up to the truck where four men waited to lift it in. We all missed our throws at one of the calves which made George Hoskins, one of the buyers, grin. But Mark Parks, the other buyer, was exasperated. He mounted a horse and rode into the herd. Whirling furiously, he threw his rope backward and caught the calf we had all missed, dragged it up to the truck where we loaded it. The expression of determination on Mark's face as he went through the process of showing us up made us all laugh; more so because he made a success of it and was roping backward to boot.

Not long ago, I was driving across the desert; it was hot and I was dreaming of a date milkshake under the palm trees farther up the road. I was cruising along at sixty when the white-haired man in a green car passed me. "Driving fast for an old codger," I said to myself as I stared after him for there was something familiar about him. I thought I recognized that old head and the set of the shoulders. I speeded up and stayed close behind hoping he would soon stop. I didn't follow far before he did stop at a service station and I drove up beside him. Sure enough it was old Mark, still going about his business as furiously as ever.

He has in his lifetime bought many thousands of head of cattle. Cattlemen and packing house operators have risked thousands of dollars on his judgment and won for Mark is one of the best judges of beef in the world. He has moved to Brawley and he seemed glad to see someone from the San Joaquin. These chance meetings are welcome nowadays because Bakersfield is no longer a little cow town but a city and we are strangers in our own home town. One can walk its streets for days and not see a familiar face.

Comedores de Carne Ajena . . .

In the old days, *zarcos,* blue-eyed men, seemed to find wives much more easily than those swains whose eyes were brown, *morenos.* This predilection on the part of the *doncellas* was not, malicious people said, due to the *zarcos* being handsomer than the *morenos.* The reason was purely economic, for *zarcos,* as everyone knew, were able, like owls, to see in the dark, and what girl would refuse a man gifted by nature to provide her with the choicest meats?

The *paisanos* in those days were *comedores de carne ajena,* eaters of other people's meat, and cattle that were slaughtered by them were roped and butchered under cover of night.

Early in my association with *paisanos* I learned that there is no better wedge than tobacco to pry open the dusty pages of memory. I had brought the old man a *regalo,* gift, and after we had sat down in the shade of the grape arbor and he had rolled a supply of cigarettes, he was ready to tell me a story.

That day Doña María was broiling meat and the odors wafted to us prompted me to ask, "Is it true that *en los días de antes,* in the old days, the *hijos del país* ate little else but meat? "Yes," he answered, "in those days we were *comedores de carne.* Meat was our mainstay because, as you know, gardens cannot be hoed on horseback. Our *milpas,* gardens, were only big enough for the women to care for. Though wheat could be threshed by driving wild horses around over the threshing floor, a *tortilla* was not ready to eat until after the meal had been ground out on a *metate.* A big appetite for *tortillas* was not considered polite by *damas de casa* who had to spend the morning on their knees grinding out the day's supply of *masa,* dough. One seldom ate more than one *tortilla* at a meal.

"In the beginning, cattle were bred for their hides and tallow. After the gringos came cattle were sold for beef and it was during this period of California history that the rancheros waxed rich and prosperous. When the big cattle companies were formed the owners frowned on the eaters of other people's meat. Henry Miller, a good judge of human nature, had a canny way of discouraging these meat eaters.

"There were a number of people, all related, who squatted on some land near one of his ranches. He knew that the numerous children were nourished on his beef, so one day when the men of the family were home he drove over and told them that if they needed beef, to kill whatever cattle were necessary for their wants but to be sure and hang the hides where his men could see them and take them to the ranch. With that he drove away, knowing full well that those people would be careful in the future to butcher only cattle that belonged to someone other than Henry Miller.

"But whatever the value of beef during the era of the great ranches, there was many a *tasajo,* piece of jerky, eaten that was not bought over the counter of a butcher shop.

"Marcos was the last *comedor de carne ajena* in the district where he was born. When the country changed from cattle ranching to agriculture, he lived on in the neighborhood and found work in the vineyards and fields to support his family. Hard times came and the occasions when his family had meat to eat were farther and farther apart. He often sighed for the great *mantanzas* when the *noqueadores* changed horses on the hour and their arms grew tired with the slaughter of so many steers and the great horns were used for miles of fencing; but cattle were seldom seen now. Marcos was of a people who believe that he who worries about the future has no faith in Providence.

"One day a yellowish heifer strayed to his place and joined his mule. She was very fat so Marcos caught and butchered her. The meat he cut in strips and hung on a line, *tasajera.* He dug a pit and put some river boulders in it. On these he built a fire of oak wood and when it had burned down he removed the coals. He then wrapped the head in wet burlap and put it in the pit. After he had covered the hole with a piece of sheet iron, he banked dirt on top to hold the heat. The next morning he uncovered the head which was done to a turn. When the last vestige of the meat had been eaten the skull was claimed by the children who used it to play *toro bravo.*

"The hoofs were toasted until the outer shell was loose enough to be removed. These, with the tripe were cooked with dry corn, *pozole.* The hide was spread out and he made holes around the edge through which he drove wooden pegs into the ground to keep it from shrinking while it dried. When it was dry he gave it to his wife to put on the floor for the children to lie on.

"The man who owned the heifer was hunting for her when he came upon the meat hanging on the line and the hide that had been put outside to air. The man, a short tempered *gringo salado,* freckled gringo, asked Marcos if the meat had been taken from the animal with the yellowish hide. Marcos said that it had and hospitably offered the man some meat to take home. The man, sure that it was his heifer's hide, went for the sheriff who came, and since Marcos denied nothing, took him before a judge. That worthy told him he must either pay for the heifer, a valuable *vaca de leche,* milk cow, or go to jail. Marcos said he had no money as work was scarce but if the gringo wished, he and his family would pay for it by laboring in the man's fields. But the gringo would have none of that and demanded money. The judge who was a member of a pioneer family loaned Marcos the money which was repaid in time by the last *comedor de carne ajena.*"

Brujas . . .

The *platica,* conversation, had turned to witchery and José Rodríguez told us this tale about *brujas*:

"There are good *brujas* and bad *brujas.* At night, when evil things are done, the good *brujas* visit others of their kind and spent the time in idle talk, doing no harm other than tiring their tongues; but the wicked *brujas* prowl through the houses and beat the sleeping occupants. When the owls hoot at night, the sleeper feels blows from an invisible hand as if in a bad dream and awakens in the morning barely able to move.

"The wicked *brujas* do many kinds of mischief. They ally themselves to robbers and thieves and aid them in their evil deeds. They lower corral bars and make breaks in the fences so that the animals can wander away and fall into the hands of men who steal horses and cattle. One must be of strong will and strive mightily to overcome their malice.

"Back in the times when cattle and horse thieves were a weary plague in the San Joaquin Valley, I was herding a bunch of cattle by day and standing guard at the corral at night. After a number of long days and sleepless nights full of alarms, I was worn out. I looked for and found a young man to relieve me. I knew he was weak willed and would be an easy mark for the *brujas* I knew were about because there is a light about them which goes on and off and I had seen it in the night.

"That evening before I went to bed, I said a prayer backward. Then in front of the corral gate I drew a circle on the ground. I then took my sash from around my waist and placed it in the middle of the circle in the form of a cross. I then took my dagger and stabbed it in the center of the cross and went to bed. Now *brujas* cannot cross this circle but will be held as though charmed until sunrise, when they assume the shapes of harmless animals.

"The next morning I knew the *brujas* had come to the corral for there they were in the shape of turkeys, huddled around the ring making shrill complaining noises that sounded like human speech. Later that day I rode in a wide circle to cut tracks and found where two horses had stood in the heavy brush. Cattle thieves had waited there for the corral bars to be lowered, but had ridden away at daylight, disappointed no doubt. From that night on I found no signs of prowling cattle thieves or of wicked *brujas* and thenceforth slept in peace."

The Cook and the Red Head . . .

Mrs. Lester Frick of Arvin, who is interested in the stories of pioneers in her locality and has done no little research on the subject, told me these stories.

One was about a shy, bashful boy who was sent to town to fetch the new cook for one of the ranches. On his arrival there he was chagrined to find that the cook was not a Chinaman as was usual in those days, but an enormous Spanish-speaking woman with a mustache. There were a number of loafers at the depot who were interested spectators as he led her to the buggy. The woman waddled up to the vehicle, put her foot on the wheel and began the struggle to get her bulk into the seat. As the red-faced boy put both hands on her broad back to help her up, she turned her head and said for all to hear, "Eef you peench me I weel tell my hoosban."

Then there was another woman who had flaming red hair. She and her husband moved to a new home some hundred miles away, taking all their chattels. The morning they started they found the family cow would not lead behind the wagon. The woman walked, leading the animal and hoping that once it got away from home it would become more tractable, but it didn't work out that way and she ended up leading the cow every step of the way. The woman walked the entire distance in her bare feet.

Canducci Castro . . .

Not long ago Canducci Castro walked into one of the largest business houses in Bakersfield. He was talking with one of the salesmen when the owner recognized his voice and came out to shake hands with him, saying, "This is Castro. Give him anything he wants for his credit is unlimited here."

It was more than half a century ago, when Canducci was a small boy, that he established his credit with this firm. His first job had been taking care of some cows for Solomon Jewett and he had earned his first money, ten dollars. He had this in his pocket, when a man driving a team hitched to a spring wagon stopped in front of his father's house.

There were two sewing machines bolted to the wagon bed. The man unloaded one and set it up and gave a demonstration by making a great number of fancy stitches. Neither of Canducci's parents spoke English, so the salesman addressed himself to the boy, saying, "Your mother needs and would like this machine and I am going to leave it here for her." The boy told him he had no money to pay for it and the man said, "Oh, don't worry about that. I will trust you." So he left the machine and drove back to Bakersfield.

That evening, a boy in a somewhat dubious state of mind called his brothers together and leading them to where the machine stood, said, "I bought this sewing machine for Mother, now chip in and help pay for it," which they did, and Canducci rode in to town a few days later and paid for it in full, proving that the salesman was a good judge of character.

The Lazy Vaquero . . .

The lazy vaquero complained constantly of his bad luck but would never make an effort to improve it so the older he grew the poorer he became.

One day a gringo who had become too rich offered him fifty pesos to go to a certain place and tell this gringo's lady luck to stop sending him riches. But the lazy man refused to go and told his wife about it. She scolded and said, "Fool, you cannot see good fortune. Now that you have missed that chance you must look for something to do, for our children are hungry." So he went to the gringo and offered to accept the job he had previously refused, but the gringo said, "You complain of your poverty but will not accept fifty pesos. I will now give you but thirty-five." The unlucky one refused the thirty-five. His wife scolded and sent him back to the gringo who said, "You would not go for the thirty-five so now I will give you but twenty-five." The vaquero accepted the twenty-five and went out and called the gringo's lady luck. A beautiful woman appeared and said, "Whether the gringo wants it or not, I will continue to send him riches." Then she left.

The lazy man thought it might be a good idea to call his own lady luck, so he called and an old witch appeared. He said, "You are ugly; the gringo's luck was beautiful."

She replied, "I could be nothing else but ugly to be the luck of a man who is offered fifty pesos and goes for twenty-five."

Wildcat . . .

The *platica* had changed from good vaquero horses to faithful dogs and Jack Glenn told this story: "I once had a dog that was about half bull. He was following me one day as I rode along a ridge toward a flat when a bob-cat came stalking out of the brush and we met on the flat. The cat stopped when he saw the dog and prepared to give battle as the mongrel lumbered up to him. Some dogs can lick a wildcat if they know how to fight, but this one didn't. He was just game.

"He charged into the 'tom' and bits of hide covered with white hair began to fly. The dog never stopped boring into the enemy and when he had fought until he was staggering and gasping for breath, I caught his hind legs and pulled him out of the battle and held him until he was rested. The cat held his ground and waited. When I let the dog go, the fight began again. After I had caught the dog and rested him a number of times, I saw that if I let it go on, the brave old dog would die so I caught him again and laid him across my saddle, mounted and rode off."

"Didn't you kill the cat?" I asked. "No," Jack answered, "I was careful not to hurt him. The old 'tom' was the better man so I left him in possession of the battle-field. When I reached the top of the ridge, I looked back and he was still there, holding his ground and glaring defiance at our retreating figures."

The Watch . . .

Mishaps and petty accidents occurred quite often in the vaqueros workday but fortunately, in most of them the range rider "lucked out" and got only a few bruises or suffered nothing worse than the laughter of the other men. Sometimes, however, the vaquero met with a serious accident, was carried away and never came back or if he did return it was to get his belongings and go somewhere else to find another way to make a living.

One morning when the crew of the Miller wagon was out on the west side of the valley an accident occurred which brought little real hurt to the victim but caused a great deal of laughter among the men. Nacho, one of the vaqueros, always wore a vest in the pockets of which he carried pencils and a notebook like a regular bossman. As though this were not enough useless weight to carry, on a trip to Buttonwillow he bought a "dollar watch" to add to his impedimenta. On his return to the wagon Lupe Valenzuela, the cook, gave him a piece of buckskin to use in lieu of a watch chain. He tied one end of the thong to the watch and the other to a buttonhole of the vest and put the watch in one of the pockets. Buckskin is the strongest of leathers so the watch was secure.

A day or two later he was riding a cold-backed gray. The other men, watching the horse travel with a hump in its back, looked at each other and grinned expectantly. The crew rode on and in a short time came to a narrow gully and jumped their horses over it. The gray jumped over too, but when he landed on the other side he kept right on jumping. At the first jump the watch dropped out of the vest pocket and swung in an arch and hit the buckeroo on the nose. The next jump, the watch gave him a black eye and as long as the horse bucked the watch swung. As we have said, buckskin is the strongest of leathers so if the horse had not quit bucking the watch "would have beat me to death" as the vaquero afterward said. He untied the buckskin thong, took the pencils out of the pockets (he had lost the notebook) and with what was left of the watch threw them as far away as he could.

The Boludo Steer . . .

Far out on the west side of the San Joaquin Valley stretch the Alkali Flats. For miles and miles the shimmering heat waves rise out of the sun-drenched ground and appear to the eye as a lake in which trees stand with their trunks covered by the cool water. This lake is an illusion. As one approaches, the water and trees disappear and one finds only burning, blistered ground. There are fertile spots here and there in this vast sea of alkali which grows a thin grass. In these scattered oases, Henry Miller established camps with pumps and troughs to water the cattle these plains supported. These desolate, weather-beaten camps were places where the men who tended the cattle never stayed for long.

The red *boludo* steer was long-legged, rawboned and very thin; one of those unmarketable animals found on every ranch in that era of crossbred cattle. This was before the Hereford had spread over the ranges of California. Season after season these steers were parted out of the herd and discarded by the cattle buyers. If they were not slaughtered for ranch consumption, they died in severe weather as old age over took them; bovine outcasts which were always a frustration to the ranch managers. They became familiar and were a joke to the vaqueros who often gave them names such as *novillo blanco,* white steer, on the Tejon Ranch. After trying for years to foist

The White Steer

him on a cattle buyer, Don Jesús got tired of seeing him culled out of the beef herds and ordered Nocho Montes to slaughter him for ranch use. Often, some vaquero, chewing on a piece of tough, stringy meat would say, "This must be a chunk of old so-and-so. It's tough enough for sole leather." Cowboys, having been fed the discards of the herd so often, called the meat they were served, *boludo,* lump jaw.

Boludo cattle have a swelling like a boil on the lower jaw, or better yet, like mumps in a human. Some cattlemen say these lumps are caused by foxtail, which, when taken into the mouth lodges there, then works through the flesh, festers and makes the enlargement. Others say the lumps are not caused by foxtail but are the result of disease and are often found in suckling calves too young to eat grass.

The red *boludo* steer was coming to water over one of the myriad trails that led to the infrequent troughs set out on the plain. Henry Miller drove into the camp, stopped his team, gazed at the steer for a few minutes then jumped to the ground. The foreman was approaching and when he was barely within hearing distance, Miller pointed to the steer and said, "When you get the time, slaughter that steer, throw the head and neck away and feed the rest of the carcass to the men." With that the cattle king stepped into the buckboard, took up the reins and drove away.

The foreman did not have the steer butchered. He had the riders doctor it and in time the animal became sound again. One year later Henry Miller drove into the camp again, stopped his team, looked out across the plain and saw the boludo. The red steer at that moment raised his head from the trough. When the foreman came within hearing distance, the cattle king wasted no words. "You are fired," he said, "I told you a year ago to slaughter

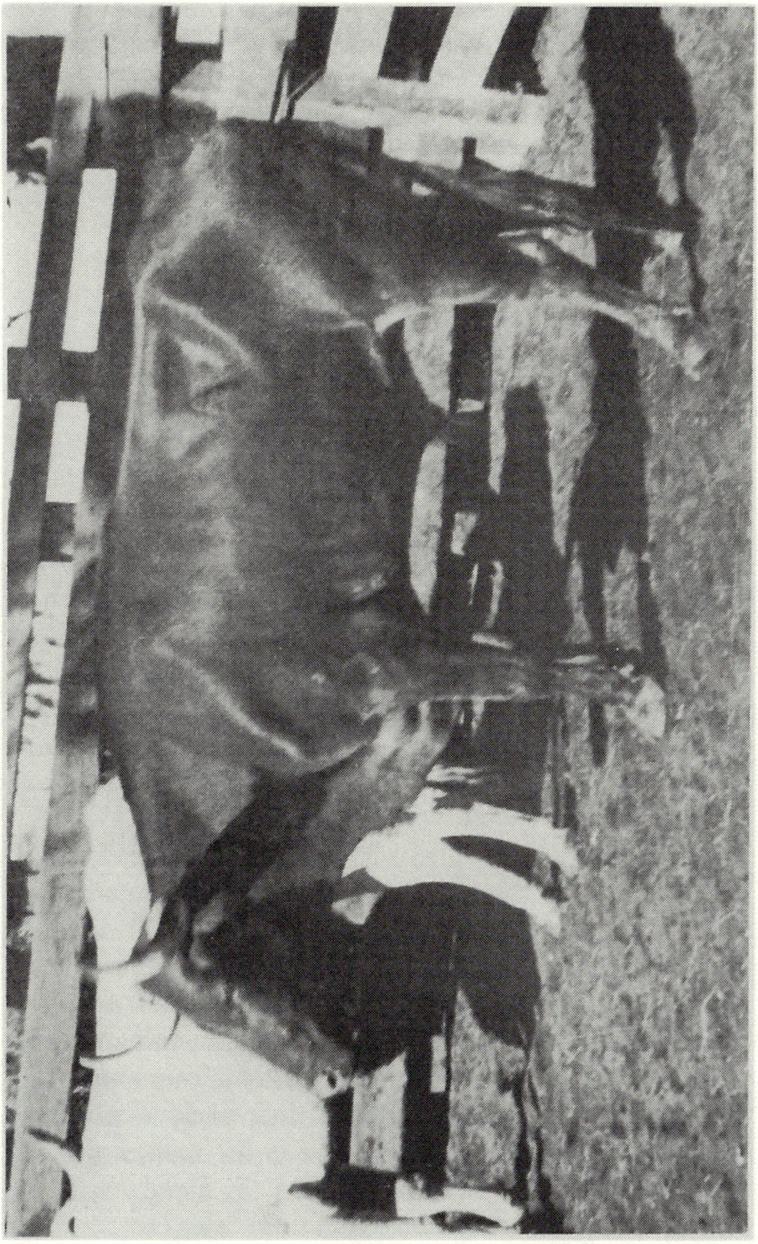

The Red Steer

that steer and you haven't done it yet." Turning around, he stepped into the buckboard and drove away. How he remembered the red steer, one of many thousand on the hundreds of cow camps he owned, such as this one on the alkali plain, is one of the marvels that made Henry Miller an outstanding figure in an age of remarkable men.

Frank . . .

No ranchero raised hay back in the last century. Frank wanted to go to a dance and went out to catch the saddle mare which had been turned out on the open range. The mare would let him get up to her rump, then she would run away and stop after going a short distance; then Frank would try again. This procedure continued until the afternoon was almost gone and the sun was low on the horizon when the mare went into a box canyon, with a despairing and weary Frank still following. The mare, seeing that she was cornered, turned and dashed back. Frank took a turn with the rope around a tree and standing on a rock, threw a loop and caught the mare. He went home and asked his father for fifty cents, but his father gave him a lecture instead of the money and Frank did not get to the dance.

Love's Labor Lost . . .

Vaqueros must of necessity spend their lives in camps. The crews are made up entirely of men, but this enforced celibacy does not prevent the range riders from exercising their genius for inventing tales of their prowess with the ladies and of how expert they were in the matter of love. The favorite topic of conversation in camps was, of course, *doncellas*.

Among the greatest braggarts and one of the most entertaining was ancient, white-haired Don Santiago. He was a frequent visitor at the vaquero camp on Tejon and was welcomed by all of us.

Though Don Toribio, the *caporal,* would say, *"Yo creo que son puras mentiras,"* he would sit with us and listen, chuckling, as long as the old man held forth around the the fire. Don Santiago had not only been an *enamorado* but a *jinete* as well. He would have us believe that no girl had ever escaped him once he had made up his mind to woo her. This little pastime was much more difficult to indulge in when he was a boy than it is today. The girls' fathers were always very strict and saw to it that the rigid, hidebound laws of *costumbre* were observed. Girls were kept behind barred windows and a youth needed courage and ingenuity to reach them.

The economic factors in the size of a family were not taken into consideration then as they are today. Since large families were looked upon as a blessing from Heaven, the girl usually had five or six strapping brothers. These brothers never failed to be *muy bravos,* very fierce, and *muy celosos con las hermanas,* very jealous of their sisters. They would take a violent dislike to a prospective suitor and seek him out to challenge him at the least suspicion that he had designs on their sister. A severe thrashing was often the price some youngsters paid for being *enamorado.*

Since Don Santiago's little *machito,* mule, needed shoes and since it was mean to shoe, he wanted us in a good humor before asking us to do that strenuous job for him. As *un favor con otro favor se paga,* we asked him for a story and this is what he told us:

"There was a girl in my town whose brothers were *tan bravos y tan celosos,* so fierce and so jealous, that none of the boys even dared to flirt with her when she promenaded in the plaza with the girls who seek husbands. One by one, the other girls were married but no one courted the girl with the fierce brothers.

"I watched her walk round and round every evening and the more she walked the more determined I became to defy her six brothers. When one is young one laughs at strife.

"Courtships begin at the plazita. From there one goes to stand in the street at the girl's window. I stood many days to convince her I was sincere in my intentions, but the window remained closed. I would never have known why if another girl had not told me that her window was kept locked by her parents.

"In those days the houses were built to surround a *corralon,* courtyard, on one side and a wall on the other. I thought, if I keep an eye on her brothers, I may find a way to solve my problem; so I dug a hole in the adobe wall and went every day to watch them. One Sunday they brought in a number of mules, as they were *arrieros,* packers, and had a trade route into the mountains. There was a red mule in the band that they were to use as a mount and evidently had just purchased. When one of the brothers mounted it the animal bucked him off. Another brother who tried it was also bucked off.

"As he was limping toward the house I was struck with an idea. Waiting until the father and the brothers were standing and staring at the mule, I knocked on the gate. When they opened it, I rushed up to the father and said, 'Señor, my father has sent orders for me to go to him at once. Could you lend me one of your animals? Any one will do; even the worst one you have, just so it is something I can ride.' Pointing to the red mule, I said, 'That one will do.' In those days, parents were taken seriously, so the girl's father said, 'We would be happy to lend you that mule but it bucks and has just thrown two of my sons who are not bad riders. If you can ride it you may take it.' I thanked him and mounted the mule.

"As I have said, I was a good rider and that *bestia* could never have thrown me. I rode for three jumps then let it throw me. I landed on my hands, striking my face and cutting my lips. Though not badly hurt, I lay there groaning. The boys carried me to the house. On the way I managed to get a quantity of blood in my mouth and when they laid me on the bed, I let it run down my chin. The mother and girl came to attend me. The mother, believing me to be seriously hurt, sent one of the boys for the *curandera* who came and gave me a *flotada,* massage. She guessed that I was not badly hurt and divined my motive for playing sick and as I paid her well, she did not let up on the *flotadas* for many days. During this time I had many opportunities to speak with the girl.

"I suppose you think it would have been simpler if I had written her a letter. I did not do this because the only man in the *pueblo* who could read and write was the padre and he was not the sort one would bother with such things. No doubt you would like to know what became of the girl." Don Santiago sighed and said, "I don't know, for having succeeded in my undertaking, I became a little too cocksure. The day I left my bed I went out and rode the red mule, without stirrups.

"The brothers from that moment became suspicious and I met cold stares from them thereafter; but I was safe and treated with courtesy while I stayed under their roof. However, I paid for my rashness after I had left their house and was saved from a thrashing by a pair of worn spurs. As you know, the rowels in a spur, after much use, become loose enough to get crosswise in the shank and lock, making a very effective hook. When a spur locks in a cinch it is because of worn rowels. Some riders who used their spurs to clinch with would often file the rowel and the inside of the shank to get more play.

"One night, long after I had ridden the red mule and had left the girl's house, I was riding down a lane. This happened long before the gringos had taken the moon and put it on a pole to light the streets' as the old people said. I was in the darkest part of the lane when I heard a whistle behind me which was answered by some one in front of me. There were many robbers in those days and I knew I would be attacked. To avoid making a target for a rawhide loop, I dismounted and removed my spurs. It is not true that every one wore a pistol in the old days; very few could afford a gun when I was a boy. I gripped a spur in each hand. The spurs were very old and the rowels were thin and sharp. Two men attacked me. Had I known they were unarmed, I would have ridden them down. As they closed in to grasp me, I used the spurs as a *punal,* poniard. I drove the five-inch rowel into one man's stomach and raked upward. The man grunted and withdrew. The other wrapped his arms around me in a hug. I placed the spur under his armpit and raked downward. He cried out and released me. As I brought the weapon down across the man's face, the rowel caught against the shank and hot blood spurted over my hands as the steel plowed furrows in his face. The man yelled, turned and fled. We had fought into a clear spot in the lane and instead of my assailants being robbers, I saw they were the girl's brothers who had been treated so roughly. Knowing they would be scarred for life and that they would seek me out, I bade goodbye to the pueblo of my birth and came into the sierras."

The next morning, after we had hog-tied the *machito* we laid a pole across his belly with the ends under his feet and lifted them in order to nail shoes on his flinty little hoofs and for good measure we put new bolts in the wheel rims of the Don's spring wagon.

We sweat a little and barked our fingers, but considered the tale he told us well worth the price.

The Mouse-Colored Mule . . .

Back in the days when California was an infant state, the desperado Vasquez and his henchman Procopio rode its back trails and people often wondered the family of Ruiz was never robbed. The three Ruiz brothers never told that this immunity from pillage had begun with the mouse-colored mule and the Chileno bit she wore when she was stolen.

The loss of the mule was not mourned so much as was the Chileno bit, for it was silver mounted and handmade to order. It was the loss of the latter that took one of the Ruiz brothers to the house of the *bruja,* witch. The witch told him that the mule had been stolen by a *zarco,* blue-eyed man, and was a three days' ride away, but in which direction the *bruja* could not tell him.

The Ruiz brothers searched the herds for the mule, but it was not to be found. Days passed, then a season and it was summer again in Southern California, when one day, as the three brothers dozed in the shade in front of the house, the mouse-colored mule passed, ridden by a blue-eyed, hawk-faced man. The brothers Ruiz were instantly awake. Without a word they walked back through the house, mounted their horses and followed the mule and rider to where he entered a courtyard and tied the mule.

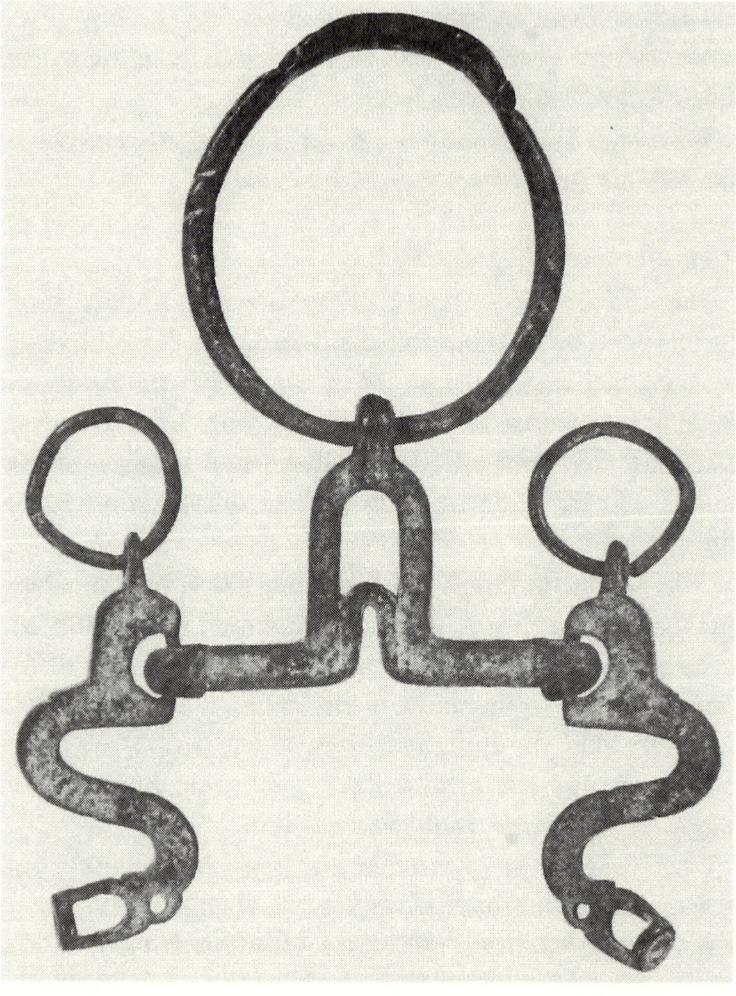

Freno Chileno (Chilian Bit)

The brothers went back home to wait for night, leaving the horses saddled. It was the custom in those days to saddle and bridle the horse on arising in the morning and leave him so all day, with the reins wound around the saddle horn just tight enough to take the slack out of the chains. It was believed that the horse would love the bit the more he had it in his mouth, and for every day the bit was left out, so much of his rein would be lost.

The horses had stood most of the day spinning the crickets in the mouthpiece. It was black night when the Ruiz brothers mounted, uncoiled and tossed out their riatas, then carefully recoiled them without a kink and laid them across the fork ready to hand. All three rode out without a word, sobered by the thought that they were going to attack a desperate man foolhardy enough to ride an animal back where he had stolen it. But the thought of recovering the Chileno bit, which had no equal in all the pueblo of Los Angeles, spurred them on. The streets, only wide enough for two men on horseback, were narrow for the throw of a riata of eighteen *brazadas*. Two of the brothers waited, loop in hand, for the return of the blue-eyed man. It was midnight before they heard the mule's hoof-beats echoing hollowly from the blank adobe walls that lined the street.

To men accustomed to horses, the distance from one approaching on a dark night can easily be judged by the sound of its movements. When the rider came abreast the brothers each threw his loop. One caught the mule, the other two the rider, pinioning his arms to his sides. Two men kept the riata tight while the third led the mule. Thus the blue-eyed man was captured and taken helpless into the Ruiz courtyard, without any one behind the walls which lined the street having the slightest notion of what had transpired. The brothers disarmed the captive and led him into a room. "Where, señor, did you get that mule that was stolen from us?" one asked.

The blue-eyed man laughed in their faces and said, "I found him one day at the end of my riata." Another brother said, "Look you, señor, if you need money we have it to give you. Even the mule we will give you if you have no mount, but the Chileno bit you cannot have, though it is the only one the mule obeys."

The blue-eyed man laughed again and asked, "How will I evade the gringos should they discover me in the pueblo, if the mule does not obey? Do you want me to be captured?" Then the brothers Ruiz knew they had captured Procopio, the lieutenant of Vasquez.

The oldest of the brothers walked out of the room and in a few moments came back to the door and said, "Come, señor, mount the mule and may you go with God, and forgive that we would have denied it to you." So Procopio rode the mouse-colored mule out of the courtyard and away from Los Angeles.

Some months later, a vaquero from one of the ranches in the hills that skirted the southern San Joaquin Valley arrived at the Casa de Ruiz in the pueblo of Los Angeles and delivered, carefully wrapped in a white buckskin, the silver mounted Chileno bit with this message, "It is not good to keep from another man that which he values most."

Fear . . .

Juan was always frank in telling his tales. "At first," he said, "I was afraid of the ground when I began to mount colts. But I was an orphan and food was scarce in my widowed grandmother's house. For many days after I had gone to work as a vaquero I was in fear, but, as the saying goes, 'Hunger is a spur that drives one on.' Of all the pains a poor man can suffer, hunger is the worst and because of it, a man will face any risk to escape it."

Fear of horses is not innate; it is acquired. Down through the ages old people have told tales of bad horses, the number of tales increasing with each generation until they make man's best friend a veritable monster. Children hear these tales and if imaginative or naturally nervous, acquire a terror before they start working with the animals and as a result are handicapped by the belief that there is death or at least broken bones to be gained from horses, though the number of persons hurt in the vaquero's work is proportionately small.

There are many tales told around the campfire of bad horses killing or injuring their riders. From these tales the vaquero learns caution and so there are many things he is careful to do or not to do. For instance: he always unbuttons the three lower buttons on his jacket to keep it from catching on the horn. He does not like to wear a belt as that too can catch on the horn. When he buys new chaps he unlaces the string that holds them together and replaces it with a thin string, passing it through the two top holes and tying a knot in each end just big enough to keep the string from slipping through the holes. These knots will slip through, however, from a hard jerk and freeing the rider if the string happens to hook over the saddle horn when his horse is bucking. He does not like to wear shoes as the heel will go through the stirrup and for that reason he wears the high heeled boot. His stirrups are tied to the stirrup leathers to keep them from turning. Whenever he has his boots halfsoled, he has the shoemaker slip the sole under the joining leather to keep it from catching in the stirrup when he withdraws his foot. He often wears two pairs of pants for when two are worn, they will not creep up and chafe his legs.

When using care the vaquero says, *"No es miedo es precaucion,* it is not fear, it is precaution."

Mules . . .

It is a curious circumstance, but a fact nevertheless, that the offspring of a mustang mother and a burro father is a much superior animal than either of its progenitors.

My good friend José Acosta, who knows mules, says that the best riders are those who learn to ride on the hurricane deck of a Spanish mule; that old vaqueros, through personal experience in a hard school, knew that a boy when learning to ride would fall off a bucking animal because of fear. The old-timers had a severe but effective method of teaching a youngster to overcome his faintheartedness. They would first catch a wild mule, saddle it, then mount the reluctant pupil on the blindfolded animal. Two men, each armed with a quirt, would ride up and take a position on each side of the snorting bronco. When the blinds were lifted and the mule sprang forward, bucking, they would keep close beside the rider. When daylight showed between the saddle and the seat of the boy's pants, the *rastreros,* as they were called, would whip the kid until he regained his seat. After a few lessons the youngster had more fear of the whip than he had of the mule and in this way learned to stay on bucking animals.

The breaking of a mule was accomplished after much hard work but after it was done the hybrids brought their riders to such a degree of skill that people would often say, "Surely that man must have the power of the devil to ride as he does."

On one *rancheria,* the Indians had a mule that could not be tamed so they drew a picture of it on the ground then took whips and beat the drawing; this was supposed to break its spirit, but whether it did or not José could not say.

Riding and managing mules has many elements of danger but when one has ridden them for any length of time he will have developed a love for these half-wild animals that give a staunch loyalty not found in breeds of loftier lineage.

A mule is catlike in its actions and has a much better sense of balance than the horse. If it happens to fall (a rare occurrence) it tries to land on its feet and usually manages to avoid being hurt. Even when being ridden for the first time a mule can run with its head doubled around to the rider's knee and keep its footing. Mules have more courage than horses; they will lead wild bulls off a mountain when a horse will fail at the task. When a mule bucks for the last time it bucks longer and harder than it has ever done before as a sort of farewell performance or maybe it wants to show its rider what it can do when it really makes an effort.

A mare mule is much fiercer and puts up a much gamer fight against man's mastery than a horse mule; but by the same token, the female is a much smarter animal than the male. Mules are gentle and obedient or vicious and stubborn; it all depends on the breaking and handling.

The gentle and obedient animals are ridden by those who take pains to listen to the words of old men and profit by their advice. A mule colt wanders over its *querencia,* home pasture, and acquires the habit of going to where the choicest grass and the sweetest water is to be found and learns to avoid the patches of sparse grass and brackish water. When a *jinete* breaks the mule by riding it over its *querencia,* he will meet with trouble. If the mule is ridden along a trail in the direction of good grass and water it will go willingly, but if it is turned toward a region it does not like, it will naturally refuse to go. Then a battle of wills takes place in which the mule will often emerge victorious. The outcome of these stubborn spells will be a hard, unresponsive mouth due to the pulling and doubling and will end in a headstrong, self-willed animal. On the other hand, if the *jinete* will lead the colt far away where it does not know the trails and has no *querencia,* the mule cannot follow habits of travel and will be gentle and obedient.

Sometimes *jinetes* will help each other when riding broncos by taking each other's lead rope and with one leading the other's animal, they manage to stay together and make headway.

A mule following a band of horses or cattle through dense undergrowth will need no guiding hand but will stay close by scent or sound and never lose them and in the end will drive them out in the open.

A mule will never be gored but once, because its memory is phenomenal, and it will never get near enough to a horned creature to be gored again as long as it lives.

Solid cheeked bits are never used on mules and the men who prefer the hybrid to the horse as a saddle animal have many reasons for their choice. There is a vast difference of opinion among riders as to why the spade bit is usually made with loose cheeks but the men who ride mules down steep mountain sides say that if one should fall the loose cheek bit will not tear its mouth as badly as will a solid cheek bit.

Don José says further that as a saddle animal the *muleto,* hinny, is vastly superior to the mule but that the latter is used more because the offspring of the female burro and a male horse is hard to breed and a horse will often forsake his band of jennies after all the hard work of engrafting the horse colt onto a burro mother. That is to say, force a colt to think himself a burro. The task of forcing a burro foal to accept a mare as mother is equally as hard and often ends in disappointment. After a burro is turned out with a band of mares he will stay with them only so long as he does not hear female mule or burro bray. In which case he will forsake his mares forever. Another way to make a burro think himself a horse is to yoke him to a filly colt and leave them together for a year or two or until they mature. This method is not as sure as switching colts, *ahijar,* and often ends in failure.

Branding Time . . .

It was branding time on Indian Creek in the year 1951. We drove up to a corral where about thirty calves were bawling in one pen while in an adjoining pen their anxious mothers peered at them through the bars. There was a jeep, a stock truck, two saddle horses, but no lasso rope to be seen. Remembering the great spring rodeos of years ago, we approached an oil drum in which a fire was burning.

We sat down to witness the most important operation on a cattle ranch. Since that distant time when the first herdsman in some Old World land turned his cattle out on the open range and fashioned some crude iron to mark them, men have branded cattle to prove their ownership.

Springtime is branding time when a man, counting the increase or decrease of his herd, knows whether he waxes richer or poorer. The brand became the coat-of-arms of New World cattle kings, a symbol of their wealth and

Corral In the Sierra Nevada

power. Without it, very likely, the cowboy's loyalty to his outfit, famous in the annals of the Americas, would never have been almost proverbial.

Four men arrived in a pickup, one an old-time vaquero who did the cutting and marking. A young man wearing a cap did the branding and another old man drove the calves into the chute afoot. The fourth man handled the winch affair which drew up the rope that held the calves legs.

The first calf to be branded went into the chute at the end of which was a triangular opening. Thinking to escape, the calf stuck his head into it and immediately a lever was clamped down to hold him there. Another lever was pulled and the side chute pressed against his chest. The chute was then turned to lay horizontally like an operating table. The calf lay there with his head and front half as in a vice. His hind legs were free but not for long. A rope was put around them and the other end was attached to a windlass with cogs and a dog. This tightened the rope until the calf was stretched out and immovable. It was then branded and ear-marked. The buckeroo acted as if he were a little ashamed of such unorthodox methods and went about his work like a machinist or plumber.

We sat there watching a procedure so radically different from the brandings of other years when cattle were not worth almost their weight in gold, and a man could show his skill and prowess with the riata amid joking and fun.

The owner of the cattle, he who wore the cap, said, as if to apologize for this departure from the old way, "Well, a calf isn't hurt like he would be if he were roped and thrown." Perhaps the thought came to him as it did to us, that more roping and riding can be seen in a Wild

West show in a city in an hour or two than in days on a modern ranch.

The last of the calves went through the chute without ever knowing the feel of the riata.

The old vaquero wiped his knife clean and put it in his pocket, saying. "Let's go to the house." He led the way in that short stiff stride that comes from wearing high-heeled boots year after year. We sat down in the kitchen and watched the old man stir up the fire. As he put the coffee pot on the stove I was sure I heard a sigh. "I get pretty tired now," he said, "Guess I'm getting old."

Yes, tired of an act on which the curtain fell long ago and tired from years spent in a calling that men have tried to preserve in the face of progress and the high price of beef.

Little Old Vaquero . . .

It is a far cry from the town of Mexicali, Mexico, to Bakersfield, the southern San Joaquin and the men who followed the chuck wagon, but the other day a little old man and a buckskin steer appeared like phantoms out of the past to link them with the present. The connection was made by a rangy muley that, luckily for a number of people, no longer had horns. The steer had jumped out of a truck and had run, head up and wild eyes rolling, through a stream of traffic down the middle of the highway from San Felipe into the business district of Mexicali. I had come to a highway boulevard stop when the steer trotted past so I turned the car to follow and see what would happen. A truck loaded with cattle passed me. There were three anxious men in the seat watching the steer outdistance them in the mass of cars. I managed to get through despite the signals and stayed close behind the buckskin.

It ran on, its nose in the air, making a futile effort to scent others of its kind in the crowded shopping district of this teeming, overpopulated town. There is no sight more pathetic than that of a bewildered, frantic animal trying to escape a crowd of humans.

The sight of a steer on the street caused people to react in several ways; some would stand and stare; others would run behind cars, but strangely, quite a number of straw-hatted men ran directly toward the animal, which would charge them half heartedly, more from fright than ferocity. There being poles and cars to dodge behind, no one was struck, much less hurt. I was right behind when the steer turned down the railroad tracks which run through town. By this time about twenty boys were in full cry after it. It would run, then turn to give battle, but the numbers were too many and it would turn tail and flee.

When the cattle truck got through to the tracks, the men jumped out, ropes in hand. I called to one, a little old man, to hop in with me and we took up the chase again.

A bumpy street follows along the railroad and we jolted over it halfway up the block where a group of boys took after the steer and turned it into the street in front of the church. I overtook it here and passed it. The little old man jumped out to rope it but the steer butted him and went on. Traffic was heavy and they drive with their horns here, but I managed to turn the car in the middle of the block. The steer's tongue was hanging out and it had slowed to a walk when I passed it the second time and the little old man got a rope on it before it butted him out of the way. Though there were buildings and many people between, the steer got the scent of the other cattle and headed for them. Too tired to run any more, he stopped, panting, as close as he could get to the others in the truck and away from the mob of yelling humans. A boy ran up and tied the rope the steer was dragging to a rung of a truck ladder. The old man caught the steer's hind legs and the bystanders helped him throw and hog tie it. With as many men lifting as could get ahold, the animal was easily loaded into a pickup and thus ended our cowboying by automobile. A resident here told me that cattle escape quite often as there are no chutes or corrals and a truck is backed to wherever a car is spotted. These cattle are old Spanish longhorn mixed with Brahma so it can be understood why they are so hard to handle.

The little old man had reminded me of old Avelino Martinez in Tehachapi and the next day I drove out to the stock yards and there, parting out cattle and using an old style California bit was my companion of the day before. When he saw me he came to greet me as one vaquero greets an old side kick, and in fact he was, for noting the "Bakersfield" on the license plate bracket, he told me he had been a vaquero for Miller and Lux under Rafael Cuen in 1918. After searching our memories and comparing notes, we named enough of the men we had worked with to find we had ridden together on the chuck wagon in the San Joaquin Valley many years ago. So here, south of the border, I find Manuel Rincon, old timer, still gamely following his calling of vaquero.

Jim Crocker . . .

Jim Crocker, according to Canducci Castro, came here from back East and for a time drifted from one place to another until he hit Gilroy where he gathered a band of sheep and some cattle. Somewhere along the way in the many ranchos and pueblos through which he traveled, he learned to speak fluent Spanish. In those days English was a strange and foreign tongue in California.

Leaving Gilroy with his livestock, he moved south along the edge of the valley to stop for a time at Temblor. Later he came to Campo del Medio in the vicinity of the tule house Tomas Castro had built as a first attempt at a permanent home after his long trek from Sonora. Don Tomás had begun to rejoice in a roof over his head and a settled existence when one day a number of armed men came to his door. They were a mixed group of Sonorans and Americans; cutthroats, all. The leader told Don Tomás he must get off the land he was on at once as some millionaires from back East were to come and take posession of all the valley. They rode away, saying they would be back to shoot him if he was there when they returned.

Castro, his dream of peace shattered, watched them ride out of sight. Just then Jim Crocker with three Americans and five Sonorans rode up. Crocker asked if some *hombres malos,* bad men, had been there. When told of what had passed, he told two of the Sonorans to give Castro their guns, saying, "Here are guns. If those men come again, one of you shoot while the other loads," and pointing to Canducci, added, "Send this boy to me and I will come and help you." He rode away, a great hero in Canducci's eyes.

Later, Castro learned that the cutthroats had been to Crocker's camp and had ordered him to move. He had answered that there were six notches on his gun already and there would be six more before he would move off the land and to start shooting whenever they felt like it. The visitors didn't shoot, but rode away and never came back again to bother Castro or Crocker.

Juan . . .

Juan always started his stories without an introduction and they were never without a trace of Indian bitterness.

"Paisanos have a long or short name, depending on the size of their father's pocketbook," he began. "The padre of my pueblo charged a *real* for such as mine while one like Maria de Los Angeles cost four. It is the usual practice to name a child for the saint on whose feast day he is born. Sometimes several saints have the same feast day. My mother had selected a sonorous, mouth-filling name for me as mothers are wont to do, but at the church when the padre asked what name I had brought, my father reached into his pocket and finding only one *real,* said bluntly and rather hurriedly, 'His name shall be Juan.' "

Juan's first job was to carry a pot of *menudo* on his head from an old woman's kitchen to her booth where she sold food. He was paid three cents a day for his work. He had been working a couple of weeks when a boy threw a rock and spilled the *menudo.* So Juan worked three months without pay before he squared accounts with the old woman for the spilled pot of stew. When he was out of debt he quit his job because he had a number of narrow escapes from having the pot hit with a rock and having the stew spilled again.

Wild Hogs . . .

The old-timers disagree about how and when the myriads of wild hogs happened to be in the southern San Joaquin Valley. Some say they were brought by the first Spaniards who settled along the hills that skirt the great valley; others say that Henry Miller was responsible for their presence. However that may be, when Jim Waldon hunted them they were fierce and wild—a dangerous and gamey fighter, fully as formidable as the boars of Europe.

The dogs which were used to hunt them in the vast swamps learned to flush them out to where their masters waited to catch and hold them. A dog on each side would hold the squealing quarry by an ear until the hunter worked his way through the tules to where the squeals led him.

The hogs were an important source of meat to the early settlers, meat which could be had for the hunting.

Don Toribio Cordova . . .

There is an old belief still current among *paisanos* to the effect that no man can come into the possession of buried treasure without the will of fate and old men tell numerous tales to prove the truth of this saying.

There is an old trail that comes over the mountains from Santa Barbara to the San Joaquin. It was used by the first padres when they turned toward the heathen tribes that peopled the great valley. The many fugitives from the mission knew and followed its many twists and turns to a refuge in the *tulares*.

Toribio Cordova was born at the foot of this trail in the little pueblo below San Emidio Ranch house and in time became famous as a *jinete* in a day when expert riders were commonplace. When I came to know him he was an old man, soft-voiced and mild-mannered. At that time he was the *caporal* on Tejon Ranch. One day we were *campiando* on the plains by Rose Station. Since Don Toribio was an interesting talker, I searched the landscape for some object to remark upon and thus start the flow of reminiscences. I pointed to a line leading down over the hills from *Mil Potreros,* Thousand Pastures, and said, "That must be the trail from Santa Barbara."

The *caporal* answered, "There is an old trail to Santa Barbara down San Emidio Canyon but the one you see leads to Pleito Rancho. I know it well because I found and lost a great fortune on that trail. I learned to ride on San Emidio and one day while yet a boy and riding with an older companion, we came to a place up Pleito Canyon where there was a circle of adobe bricks. The circle was too small and too far away from the ranch building to have been a corral. It stood on a shelf that jutted out from the side of a mountain and lay in a sort of depression. It could not be seen unless a person stood over it. Cattle had died there on the shelf and the grass had grown thick and tall as it does wherever dead bodies enrich the soil.

"The circle of adobes could be seen from a distance in winter when the ground was bare and clear of grass. My companion said that surely there must be something buried there and we decided to return to the spot the next day and dig, but mañana never comes. The next day we were sent in another direction and when we returned to that vicinity some days later we forgot to take a shovel and left the place without digging.

"I was a young man before I saw the place again. A friend and I rode up to it one cold day and as we did years before, my companion and I decided to return later and dig. But holes cannot be dug from horseback and the shovel was left in the shed at the ranch when we rode away the next day and we never went back to the circle of adobes.

"More years passed. It was rodeo time in the San Joaquin and the vaqueros were camped at Pleito. One night when the crew was lounging about on the ground, a light appeared up on the hillside. I immediately thought of the circle of adobes and buried treasure. I knew deep grass covered the spot at that time of year so I got an old Winchester rifle and using a wheel as a prop, I sighted the rifle on the light, tied it to the reach and went to bed. In the morning we were sent out on a long circle and when we got back the boss said we could rest and shoe our horses the next day. The cook hadn't bothered to remove the rifle and so the next day a vaquero rode up the hill to the circle of adobes while I directed him by sighting with the rifle. When he had located the spot he motioned for me to go up to where he waited. When I got there he pointed to an empty hole within the circle of adobes. Parts of a chest lay scattered about. The boards were of a hard black wood and were covered with strange markings or symbols. The chest had been placed in the hole and boulders packed around it, forming a solid wall in the empty hole."

I said to Don Toribio, "You were close to riches, why didn't you dig sooner?" He replied, "One can never have treasure which is not intended for him no matter how close he may come to it, and that was surely not intended for me."

I asked him what he thought the symbols or markings meant and he answered, *"Pues, quien sabe,* well, who knows. *Hay algunos cuentos,* there are some tales, of Chilenos coming over this trail from a ship at Santa Barbara. The chest may have come from across the seas from some strange land."

Jealousy . . .

Francisco, Teodoro and I were riding away from camp on our way to Carneros. Every few minutes the newly wedded Teodoro would turn and look anxiously back at the closed door of the little house where he had left his bride. He seemed to have but little faith in the locks that protected her.

Francisco was Teodoro's cousin and so could speak familiarly even when a question of family was involved. Francisco remarked, *"Los celosos son castigados de Dios,* the jealous are punished by God," and added, "There is a horse-faced woman who appears and follows jealous men." As we plodded along, our horses abreast, the old vaquero told us this tale:

Juan was a very jealous man. Whenever he left the house he would make sure his wife had locked herself in and he would tell her not to open to anyone until he returned. This trait of character did not prevent him, however, from having an eye for any girl with a well turned ankle or rounded hips. One moonlight night he was riding to the pueblo when a woman appeared walking on the trail ahead. She was tall and well formed and had long black hair which hung down her back. As Juan was *muy enamorado,* he spurred his horse forward to overtake her. He got his horse into a run but could not come abreast of her, though she never moved out of a walk. Juan called to her to stop, but the woman walked on, her head bent down. He followed her, paying her compliments and saying flattering things.

He had gone many miles when, as the moon came from behind a cloud, the woman turned and raised her head. Her head was a horse's skull and the long hair Juan had admired was a mane. Juan's horse snorted, spun around and ran. After some distance had been covered, Juan turned and saw that the woman, her teeth bared in a grin, was a length behind him, and no matter how fast the horse went she never lost ground. She followed him to the gate of his house where he fell off the horse, screaming.

His wife heard him and put him to bed where he laid for many days, burning with fever.

Whether this experience cured the *celoso,* jealous one, or not, the old vaquero never told us.

Vicente's Story . . .

The *dicho,* proverb, which lends grace and adornment to the Spanish language, probably had its beginning in the ages long past when the Arab story tellers crossed the Strait into Spain with Tuerto's desert horsemen.

The *dicho* has imparted a peculiar fatalism to those who speak the tongue; it expresses a certain resignation to one's destiny. Though the *dicho* is often heard in the vaquero's *platica,* I am sure that very few of those that I knew had ever read Cervantes. Nevertheless, the old soldiers' proverbs had come over the sea, crossed the continent and up the Pacific coast to enrich the *Californianos'* language. *Paisanos* esteem the *dicho* as one of the most valuable legacies left us by mother Spain, which brings to mind Vicente and a story he told.

Vicente had finished shoeing the forefeet of the gray horse and moved back to his hindquarters. He put his left hand against the horse's hip, leaned down and grasped the hind foot with his right hand. The animal jerked his foot away and kicked sideways, striking the vaquero on the thigh. He hobbled away, and after limping about in a circle until the pain was relieved sufficiently for him to stand firmly on both feet, he went back to try picking up the horse's foot again. A number of men were watching; one of them, a mule skinner, said, "Look out, he'll kick you again." Now, a vaquero does not like to have anyone say "look out" or "be careful" or to give other warnings because he is already "looking out" and warnings make him nervous and if a man is nervous he scares the horse. Vicente frowned and answered, *"Eso queda con Dios,* that remains with God," and went about his work. This time the gray didn't kick and the vaquero finished the job.

That evening he groaned when he sat down by the fire in the bunk house and the talk turned to the narrow escapes from injury that occur frequently in the vaquero's life. Vicente said that no matter how often or how close death comes, "No one chooses the place where he is to die," and told us this story.

"Luis was a good *jinete* and the strongest man in my town and entirely ignorant of the power in his arms and wrists. One day he was sitting in a cantina when the town's *borrachito,* little drunkard, entered and seeing Luis, begged for a drink. When the vaquero refused his request, the drunkard abused him. Luis endured the vile language for awhile, then catching the pest around the waist, carried him to a corner and dropped him where he lay without moving. Some time later it was found that he was dead. Luis, accused of causing his death, was jailed and in the course of time was tried and sentenced to death. His family and that of the *alcalde* were good friends so that official was assailed with pleas to save the culprit's life. To all these requests the alcalde answered pompously, 'The law is the law.' The day that Luis was to stand with his back to a wall and meet his death drew nearer and nearer without the *alcalde* finding a way to set him free.

"So matters stood until a man was brought into town on a litter. He was mortally hurt and died within a few hours. It was the old, old story of a bad horse and a luckless rider. The horse that caused the man's death had also injured other men. He would buck in a spin until he fell.

His rider, if he stayed until dizziness overcame the horse, would be pinned under him and seriously hurt. The townspeople gathered around, lamenting that so vicious a horse could not be conquered before more men were killed, all ignorant of the fact that it is man who makes a horse vicious. The *alcalde,* like all good *politicos* the world over, saw in all this an opportunity to gain his own ends. He went before the people and when the right moment came, said, 'Luis could ride this horse but he is in jail,' whereupon they clamored for Luis.

"So Luis was brought before the *alcalde* who told him that if he rode the horse he would gain his liberty. The vaquero answered, 'bring on the horse, for there is a remedy for everything but death.' When the horse was led into town, they offered Luis a new saddle but he would have none of it and sent for his own which was rigged according to his own ideas. He saddled the horse and mounted. When he had removed the blinds, he took his hat in his hand. The horse leaped into his bucking spin and Luis let him go for awhile, but before the animal became dizzy, the vaquero hit him with his hat covering his eye. The horse shied away, came out of his spin and straightened into a run. When he had tired himself out, miles away from town, Luis turned him around and rode him back before the *alcalde,* who said, 'You surely must have faith in Providence to ride that horse back here after he had taken you so far away. My advice to you is to put as much distance between this town and yourself as fast as you can.' "

Vaquero Grit . . .

Characteristically, the vaquero laughed at his narrow escapes from injury, making as little of them as possible.

He accepted the hurts, blows and hardships he suffered in his daily life as occupational hazards and shrugged off the cold breath of pain as a duck sheds water.

The gangling sorrel horse was tied to the snubbing post and as the vaquero walked by, the animal suddenly whirled and lashed out with both feet. The hoofs grazed the man's chest close enough to tear his shirt. The vaquero, a little pale, turned to the men who stood by and had seen his narrow escape and said, "Whew! the Old Man up above had his arms around me that time." One of the men answered, "A miss is as good as a mile," and the incident was closed.

Memories of Tejon Ranch...

A native Indian was the first friend I made on the Tejon. His name was Nepomuseno but was called Cheno. The men were very unfriendly, so instead of mixing with them I put in my free time lying on my bed playing solitaire or reading the one book I owned, *Robinson Crusoe.* During those lonely months I read it over and over again. I still know most of it by heart and am grateful for the many pleasant hours it gave me. In time I was to become the friend of all the Tejoneños, but that is another story.

One day we were in camp and Cheno, who was ill with a fever, came to me and said, *"Deme una medecina, amigo."* Luckily I had a quinine preparation left from a bout with the grippe. I gave him four tablets and at intervals repeated the dosage. I sat with him through the night and by morning the fever was gone and he was up and about in a few days. From that time on he always rode with me and though he scolded me when my stupidity was more than he could bear, he watched over me and taught me many things.

From Cheno I learned that a good vaquero need not be a bucking horse rider, for he, Cheno, could never stay on a horse after the first buck no matter how mild the jump. The *mayordomos,* however, always gave him colts to break, because he was a man of infinite patience with young horses and he turned out some of the best working horses on the ranch; well reined and well trained in every way. He was also a good roper.

Cheno has been gone a number of years now, but my old friend was brought back to my memory when I saw a little grayish mule at the old fair grounds and his owner, a lean, wiry devil-may-care young fellow who must be as active as a cat for he stakes his life on his nimbleness and speed in earning his living; he fights bulls in rodeos. The men who risk life and limb riding bulls in rodeos depend on the bullfighter to save them from being gored when they are thrown and are helpless on the ground. This young man and his mule reminded me of another tall, lean, devil-may-care man and another mule of some thirty years ago.

We were breaking camp at Alamos canyon. Cheno and I had caught the mules that pulled the chuck wagon, the cook being sick at the time. We were hitching them up when one of them took a stubborn notion and refused to get over against the tongue but pointed its ears forward in a semblance of terror and backed away rapidly, snorting and dragging Cheno who was holding onto its halter chain. Some distance from the wagon it jerked away and ran in a circle back to the mule I had tied to a wheel. A tall young man came to help. He caught the runaway and said, "Hold it for me to mount." Cheno got hold of its ears and covered its eyes. The tall man mounted and rode the mule bareback. It bucked a long time and after its rider had dismounted, it was glad to get into place and be hitched to the wagon.

I learned that the man who had ridden the mule was Mariano Sanchez who had come from Rancho La Liebre, a part of Tejon. He was different from the other men. While they were stolid and unfriendly, he always had a grin on his face. He was the first man to ever call me "*compadre*." He may have done this because I was also tall and skinny, a typical Sonoran, or perhaps it was because he had Yaqui blood in his veins; Yaquis are courteous as are all brave men over the world, and call other men "*compadre*." They borrowed from their enemy, the *yordi,* white, this term of address which can express more esteem and respectful deference than any other word in the Spanish tongue. I was young then and very green and made many blunders but he never laughed at me. I remember the time I put the saddle with the twenty-six inch *tapaderas* on the bay horse that liked to kick and buck. For a while he tried to get rid of the *tapaderas* by kicking at them. When one of them hit him in the nose, he bucked around the corral trying to dislodge the offending equipment. When I caught him again he wouldn't let me get on. All the men had left but Mariano who had foreseen I would have trouble and had waited for me. He helped me mount and rode with me until the horse warmed up and got the hump out of his back.

Through the years after I left Tejon, I met Mariano now and then. He worked for Cliff Burrell and other men who owned strings of horses where he was in his element and made a good hand.

Once I was riding in a parade in Bakersfield when some one on a wild colt rode up beside me. I turned to see who would be riding a green horse on the streets and found it was my old friend and guardian, Mariano. He told me it was the second time the horse had been ridden; but to ride him in a parade!

That was the last time I was to see the compadre of my youth. I will always remember him as a man who was afraid of nothing and serenely indifferent to whatever consequences his actions might bring.

The tall young man who owns the grayish mule and fights bulls in rodeos is Mariano's son, Joaquin Sanchez.

Nacho . . .

The first time I ever saw vaqueros in action was in a corral on the V7 ranch some forty five years ago. They were branding cattle and I had the great good fortune to see them at their tasks; the last *Californianos* who took great pride not in what they did, but in how they did it. Among them was Theodoro "Cholo" Valenzuela. The dapper little Cholo, who wore his Poblano hat at a jaunty angle, attracted me with his skill in handling the reata. He would make an enormous loop and throw it without whirling. The top of the loop, coming from behind would hit the calf at the base of the neck and when the calf jumped and raised his end, the loop would shrink around the hind legs as if an invisible hand had guided it.

I remember that all but one of the vaqueros wore black angora chaps. The man who wore plain leather had on a pair of fringed, tight-legged, to my youthful eye, abominations; not at all pretty.

In the half century or so that Mr. Albert Goode has run the V7 brand, he has employed many good vaqueros and one of the best of the men who rode for him was Cholo. In a conversation with Mr. Goode, I mentioned the old wizard with the *reata* and he told me that he once had a bunch of cattle in a feed lot close to the residential section of town. In this bunch was a big steer that would jump out of the corral and roam through the lawns and flower beds in the neighborhood. It was quite a problem because the animal always chose the cover of night to take his leave. After a number of complaints from the residents, Mr. Goode drove out to the ranch and brought in his best vaquero and explained the situation to him. That evening old Cholo saddled his horse, rode out to the corral and waited. It was midnight and pitch dark when the steer jumped out of the pen. The vaquero must have been watching the skyline to have been able to see the leaping animal, but be that as it may, he roped and hog tied the brute and left him until morning. For the next three nights the steer tried to escape but Cholo would be waiting and would rope and tie it until it was cured of jumping fences.

Many years after I had met Cholo on the V7, I was in the stable business on the outskirts of Bakersfield. One day a customer of mine brought in a big Roman-nosed horse with four white stockings and a blazed face. He looked too heavy for a saddle horse and resembled a small Clydesdale to some extent. He was not as heavy as a Clydesdale and was the wrong color, being a light sorrel. Some days later I had some work to do so I saddled him and mounted. I had not ridden far before I found I was astride one of the best reined horses I had ever ridden. I had been told that he was a Cuyama horse, so, curious to know who had "broke" him, I drove to where Cholo, then an old man, could be found. He accepted my invitation to come and see the horse and when he saw him he smiled and said, "That is old Nacho. I rode him when he was a colt in Cuyama." Well, no wonder he was such a good horse. A few days ago I inquired of June Uhalt if old Nacho was still alive and owned by her father, Frank

Smoot. She told me the horse was used last summer and can yet, despite his years, show the world that well-reined horses were the pride of Cholo Valenzuela.

Trees . . .

No home in early California was complete without a fig tree, a grapevine and an olive tree in the yard. The first seedlings were brought from Sonora by the colonists. Besides giving shade they produced wine and food. The colonists also brought the oleander which was planted in the plaza, not so much for its beauty and shade but because it was the only plant that would grow. A burro will eat and can digest anything and in working its miracle of survival, disregards the symmetry of man-made gardens and devours all green things but oleanders. He gives this poisonous plant a wide berth.

Owners of burros were never arrested for letting their animals go astray; but the burro caught browsing in the plaza was impounded until the owner bailed him out.

More Memories of Tejon . . .

Juan Gomez had only one leg, so he used a crutch to hold himself upright while he put his foot in the stirrup to mount his horse. He laid the saddle strings across the cantle and tied them to the fork to make a brace for the stub that remained of what had once been his lower limb. But once in the saddle, with the crutch under his arm, he was the equal of any of the hard riding vaqueros that rode the Tejon range.

It was in far off San Luis Obispo about 1912, many miles from his home ranch in a corner of the San Joaquin Valley, and years before I met him in the flesh, that I first heard of Juan Gomez. The *platica,* talk, around the fire had turned to *vaqueros de nota,* and Antonio Moss, the *mayordomo,* told of the rider on Tejon, who, though he had but one leg rode colts so wild he had to blindfold them before he could mount. Since that time in distant and widely separted parts of the state, I have heard Juan's name mentioned whenever men spoke of good vaqueros.

Eduardo Valdez, he of the light hand on the bridle rein, who rode the best string of horses on the ranch and took such good care of his equipment that his saddle is still good after thirty years of hard usage, had passed on, one vaquero less among the few who remained in the great valley. Urrea, Chavarria and I joined the little group that accompanied his remains to their final resting place. Perhaps the paucity of old white heads bared during the service brought the grim thought to the few there that they were not far from the end of the rail. When we had left the *campo santo,* cemetery, we paused for those few last words which are always said before a leave-taking, and we cast about for something to wipe away the dark mood that lay upon us. Lupe Gomez joined us, and Chavarria went back in retrospect to his youth on the ranch long ago. He said, "Here is a man who rode with me when we were young. Let him tell you of his brother Juan who lost his leg when he was a small boy yet learned to be a vaquero."

We listened eargerly while Lupe told about Juan, who, despite his handicap and many setbacks undertook to follow an occupation in which many strong, able-bodied men have been found wanting. But he adapted himself so well that he was known from one end of California to the other. When Juan first started to ride he would tie his crutch to the saddle after he was mounted. He followed this practice until the day he was given a little black mule to ride. This little mule was hard to mount, for the moment one put his foot in the stirrup and swung on, the mule would break into a run. Most of the men who had to ride it would tie it to a post, mount, then untie the rope; the mule was perfectly gentle otherwise. One day Juan tied it to a post, mounted and tied his crutch to the saddle. Before he could untie the rope, the animal, eager to be off, sat back and broke it. The sudden release threw the mule over backwards but Juan fell clear. The mule, active as a cat, regained its feet almost instantly and ran away with the crutch. Juan never tied it to the saddle again but carried it under his arm. No matter how far or how long the ride or how tired his arm became, from that time on, whenever he was bucked off or the horse fell, his wooden member went with him.

A good roper, his skill in handling rope, crutch and horse when catching wild cattle was a marvel of dexterity. There were many little mules used on Tejon in those days. Juan was one day riding another little black mule named *Garrapata,* Woodtick, behind a bunch of cattle. He and the other men were roping at the animals in practice. Juan had heard of the "tie hard and fast" method. Taking a sudden notion, he tied his rope to the horn and put a loop on a big red steer catching it around the chest, the rope between the forelegs. The steer pulled the saddle, which was loose, over the mule's withers onto its neck and the cinch pulled its forelegs up to its nose. The mule fell and when the saddle was stripped off got up and ran off, braying in protest. The steer dragged the saddle through the herd before one of the other vaqueros could rope it and rescue the much scuffed piece of equipment. José María Valasquez, the *caporal,* rode up just then, crying, "No! no! with *reata* tied! no! *muchachos,* no." He was so upset because in California a reata just is never tied to the saddle.

Pedro Rivera or Lupe would ride a colt once or twice then turn it over to Juan who would finish breaking it. Neither Pedro or Lupe cared to ride it again after the horse was used to Juan as it would jump or perhaps buck when a leg was thrown over its back.

One day Juan was riding on a roundup with two other vaqueros when a *canelo,* roan, horse one of the men was riding started to buck and after being thrown three times in succession, the man would not mount the horse again. They were a good distance from the rodeo ground and the hour was growing late. As the unfortunate rider continued to refuse the roan, Juan traded horses with him and they then rode to where the vaqueros were holding the cattle.

Pablo Apadaca, José María Velasquez, Lupe Gomez and another man were night-herding a bunch of cattle. There were a number of *ajeno,* other people's cattle in the herd. The moon was full and night almost as bright as day. The men could see well enough to read brands, so every little while one of them would part out a steer, an *ajeno* of course, for the others to rope for practice. One of them roped a steer by the head and Lupe caught the hind legs. His horse, which was green at this work, shied away as Lupe took his turns around the horn. The horse spun in a circle. Lupe caught his finger between the rope and the horn and it was cut off. After he was untangled from the horse he gathered cobwebs and applied them to the wound to stop the bleeding. He never went to a doctor as the vaqueros cured him. They would chew tobacco until the cud was soft and mushy when it was placed on the wound. Lupe says that tobacco has great medicinal value because, after a short course of this treatment, the stub was healed.

Much of the riding in the roughest terrain on Tejon was done on wiry little mules. One day Ignacio Montes, Leonardo and Francisco Chavarria, José Moreno, José María Velasquez and Emiliano, all mounted on mules, rode away to fan out in a circle and bring in the cattle. Each followed a ridge up to its crest, then turned back, driving all the cattle within their circle before them to the rodeo ground where the *puestos,* posted men, waited to stop and hold them until all had arirved. Then the work of parting out beef would begin. The vaqueros were forced to ride hard and fast to keep the cattle from turning back. Many times a *reata* or a hat was snatched off by a branch to cause the rider to halt and lose sight of the animal he was driving. The mules were so small and the trails so worn that the stirrups at times would drag along the ground. Whether it was due to the mules being better in the mounains than horses or whether the vaqueros rode harder and faster than usual, when Mr. Lopez arrived he found many steers and unmarked bulls that had been missing from the herd for a number of years. The first thing he asked was, "Who were the men who brought down all those wild cattle?", and Velasquez answered, "The men who rode the mules." Mr. Lopez replied, *"Buenas! las mulitas buenas! Buenisimas!"*

After Lupe's tale was ended, Urrea told us of Don Ventura Cuen who was so rich that besides throwing gold pieces at the dogs when they angered him, used paper currency to light cigarettes. Don Ventura owned the most beautiful horses in this end of the valley and rode into town with his men to see Mateo Palmas fight bulls in the plaza here which was located on what is now Chester Avenue between the city hall and the courthouse. Don

Ventura owned the old Panama Ranch which was a *pueblito* before Bakersfield came into being.

So many cattle roamed the valley that when men drove a herd north to San Francisco it was necessary to send a crew ahead of the drive to push the cattle out of the way so the herd could pass through. A cattleman could start at the south end of the valley and if he were not careful he would arrive in San Francisco with a herd many times larger than the one he started with. Wild horses were so plentiful that one would be roped and mounted when ever the need for a fresh horse arose.

When Chavarria's turn came to tell of the old days he recounted his first try at running away from home. That morning he had left his home at Elizabeth Lake, *Laguna de Chico Lopez,* and followed the man with the pack horse who had passed his home early that morning. The man had fed him at noon and when they reached Willow Springs they made camp together and were sitting by the fire when Chavarria's father arrived. Without saying a word his parent put him on the horse *anancas,* on behind, and turned back toward the lake. After traveling some distance his father asked, "Why did you run away from home?" and the boy answered, "Because I didn't want to herd sheep. That ram is *loco;* when I go to turn them he butts me." He never had to herd the flock again.

He remembered the incident because Vasquez, the bandit, arrived that night and took a bay horse this time instead of his usual gray and Chavez, his henchman, took a sorrel. They left a jaded blue roan and a black in exchange. His father put his iron on these two and when he left to go to die in Sonora he drove these two back over the Camino. Chavarria told us of *Naranjo* and *Pico Blanco,* the two best horses Luis Lopez owned, and of *Tambor,* Drum, named thus because his belly stuck out and was round like a *tambor*. The horse had been ridden by an Englishman so Don Luis put on the Spanish bit he had brought from Los Angeles and the black saddle with the long taps and set about the task of restoring life and feeling to a ruined mouth.

Charlie Cohn, the cattle buyer, was there the day Don Luis rode *Tambor* into the corral with the maddened bull. In the struggle that followed his entry, the bull pinned the horse against the bar but before he could gore, "the best vaquero that ever lived" put spurs to his horse's flanks and jumped him over the bull and out of danger. Cohn, who had watched the whole incident, said, "Put that horse on the train with the cattle, no matter what price. That horse must be mine."

One unique experience in the life of Kern County vaqueros was the great elk roundup and for a long time was the subject of *platicas* around the fire. In one picture taken of this great rodeo was a big grey horse pointing the elk herd. This horse became famous because of the picture and because he was the only horse with speed and "bottom" enough to stay on the point and turn a herd of running elk. This mustang strayed away and ended up in Santa Barbara Canyon running with the Reyes horses.

When Chavarria went to work for Senator Wagy, he told his boss he would capture the horse. Doc Myers and he relayed the runaway and finally Doc, who had a good *riata,* got in roping distance, threw and missed. Chavarria meanwhile had closed in on the horse; he cast his *lazo* and caught the grey. After this fine throw, as long as he worked for Mr. Wagy, he never roped again but kept the *reata* coiled, unused. He did not want to chance losing the reputation he had made as a roper with that one throw on the grey horse.

Henry Miller . . .

Many tales are told of Miller's parsimonious nature; of how he examined the garbage cans to see if the cooks were wasting food and of his complaining to them that the potato peelings were too thick. But regardless of what was said of him by malicious people, Henry Miller took good care of his horses. R. L. Hughett, who in his boyhood worked for the cattle king, says that the men who drove the teams which were used in the fields had orders to stop them every hour on the hour and give them a rest and raise their collars so that the breeze could cool their shoulders. The teams had a drink in midmorning and midafternoon.

Another tale about Miller goes like this: Miller's driver would stop at a gate, hand the reins to the cattle king, climb down and open it. After Miller drove through the driver would close the gate, resume his seat, take the reins and drive on. This was the routine and had gone on for some months when one day they came to a gate made of new lumber. As the driver started to hand the reins to Miller, the latter said, "I'll open this one," and jumped to the ground. He walked up to the gate, inspected it for a few minutes, then walked back to the wagon, took an axe out of it and chopped the gate into several pieces. He ordered the driver to drive back to the ranch house where he fired the foreman and carpenter for making a gate out of finished lumber.

Mae Arnold . . .

Had we not been friends for many years or had I not heard numerous tales of her skill and daring, the road we traveled to reach her place would have given me some insight into the rugged character of the lady who lives at the end of its fifteen miles of little more than wagon tracks. Only a dauntless soul would choose to live in such a location. The dirt road branches off from the black-top that runs through Caliente Canyon, then turns to follow the twisting bed of Indian Creek. A mile or two from Lorraine we dropped into second gear and except for the time we crawled in low over some "rough spots," we stayed in that gear to the end of our journey. But we didn't mind the creeping progress or the bumps and jolts for we were on our way to visit Kern County's grand old cowgirl, Mae Arnold, a horsewoman who has never taken a back seat to any man when hell-for-leather riding was the order of the day.

Old-timers who have ridden after cattle with Mae Arnold say that it took determination and much spurring to stay in her dust in rough mountain country.

We opened and closed three gates before we came to the house and corrals, situated on a little mesa that projects out from the side of a mountain. There are fruit trees and one can imagine the seedlings, wrapped in a gunny sack brought in a wagon or on a pack horse, and the stops made along the way to take water out of the creek to wet the sack lest the young trees die. There is a phrase in Spanish that describes this toil, *"con que sacrificios"* and means that something has been accomplished by endless lifelong toil and effort such as this little ranch house and orchard on Indian Creek.

Mae welcomed us with the information that the ranch country's symbol of hospitality, the "pot of beans," was on the stove and to come in and share it for it was meal time and a plate must be set for one's guests.

So we sat and ate and yarned and left Mae to do the talking for her picturesque way of expressing herself is delightful. She told us of having been gathering her cattle this summer. She was a good distance from home and the August day had turned hot and she became sick and dizzy from lack of water. She was on the verge of collapse when she came to a damp spot in a creek bed. So she dug into the wet sand hoping to find water. She scratched a long time and finally enough moisture seeped into the bottom of the hole to make a swallow. She is sure she would never have got home without that drink of muddy water because she can't stand heat and thirst as well as she could when she was younger.

Mae says you must never grow careless around a bull because they are always dangerous. She told us of the fine Holstein bull which had become so mean that it had been sold to the butcher for twenty five dollars. She was helping move the herd in which the bull had been placed for the trip to market. They came to a paved road where the bull balked. Mae took her *reata,* whirled it around her head a couple of times and brought it down on the bull's rump. The bull, instead of going forward, turned and charged her. The horse, unable to make headway on the wet slippery pavement, was caught and lifted clear of the highway. He fell on his side with Mae underneath and she says, "The bull tried to hook me but couldn't reach me under the horse but he licked my face until it was clean and nice and combed my hair real pretty. His horns had been sawed off to about two inch stubs and with these he butted the horse until he sawed a six-dollar cinch in two." Mae suffered a broken hip and believes it will bother her when she is older. She is a mere seventy now.

There was another time when she used her *reata* in the same way and again met disaster. It was snowing heavily one day when she rode out to see how her cattle were weathering the storm and one of the dogs followed her. She had ridden through some thick brush when they met a wildcat. The dog, part fox terrier, jumped the cat, which turned to fight. He had a death hold on the dog and was choking the life out of it when Mae went to the aid of the terrier. She took down her rope and whirling it to get momentum, brought it down on the cat's back and she says: "The cat spit out the dog and jumped at me, missed, and clawed the horse which lit into bucking. I lost my hat in the creek and almost got bucked off. It took three rocks to kill that cat."

We expressed our regret at her absence from our fiesta September 16. She said she was sorry she had missed it because the last Mexican celebration she had attended had been back in the nineties.

In those days *El Diez y Seis* was observed in the old way. The vaqueros took a rooster and tied, as securely as only a vaquero can tie, a buckskin bag containing a ten-dollar gold piece to the rooster's legs. Then they buried the bird in the sand with the head and neck above ground. Next each man would start at a mark and put his horse into a run. Hooking a spur to the cantle or horn he would hang down the horses side with the back of his hand dragging the ground, his fingers spread in a V in which he was supposed to catch the rooster by the neck and pull it from its tomb. One after another the vaqueros tried but the horses would not run true and the fowl remained buried.

Mae was there, riding a side saddle and since as she says, "I was young then and could wear an eighteen corset," put her horse into a run, hung down and pulled the rooster out of the ground. Pulling herself back into the saddle, she kept her horse dodging the men who rode after her to take the rooster away. As she avoided one after another, she chewed on the buckskin bag until she got the gold piece out. Holding it in her mouth, she used the rooster as a club to hit the vaqueros over the head whenever one made a clutch at it. There was little more than legs left when the vaqueros gave up and left her to ride off with the honors and the ten-dollar gold piece.

Livery Stable . . .

Though it suffered a lingering death the last livery stable in Bakersfield survived until the twenties, when the old C and C Stables which stood on the corner of Nineteenth and N Streets, after a hard struggle finally gave up the ghost. Toward the last an enormous Winton, a Dodge and a Model T Ford had usurped the space where buggies and coaches were once stored.

Old-timers will remember the owner, genial old John Ansel, and Ike the stable man, who always had a chew of tobacco in his mouth. Ike was tongue tied and this made his speech worse than it otherwise would have been. When the cowboys were broke, Ansel allowed them to sleep in the hayloft and let them store their bedrolls in the harness room. Then too, John could be touched for a little money to eat on until some cattleman should come in and hire the buckeroo who had "blown in his wad." And if worse came to worst, a rider could hock his saddle to the stable owner and let his boss redeem it when the cowboy went to work. Buckeroos, when in town, would head first for the old barn where in its horsey atmosphere they exchanged news and gossip from the different ranches.

Things were discussed in the order of their importance. One of the first questions a rider would ask another was, "How is the grub?", the next, "What kind of horses are used on the ranch?" "How was the boss to work for?" and, "What kind of country is there to ride in?"

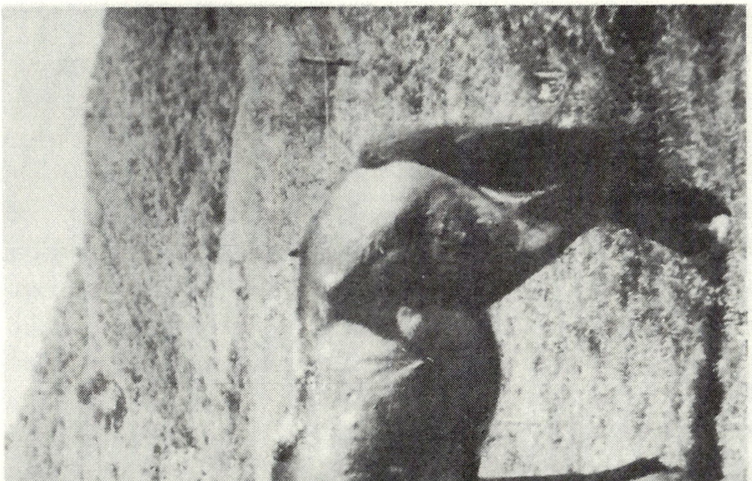

Buckskin Horse With the Honorable Marks Of Service

Chuck Wagon . . .

Strangely enough, the wages and hours worked seemed to be of minor importance. The long work day from before daylight until after dark was taken as a matter of course.

Each ranch had characteristics peculiarly its own, an atmosphere distinct from any other ranch. The Tejon was known for long hours, rough country and good but mean horses. They were good in the sense that they were well reined, tough and surefooted; mean because they all bucked to a certain degree and all had *mañas.* This word is used by the California vaquero in different ways and can be interpreted as a vice or bad habit in a horse, or as skill or cunning in a *Cristiano.* Both could be full of *mañoso.* So, if a horse had a *maña,* vice, by the same token a man must have a *maña,* skill or cunning, to be able to handle the brute.

San Emidio was rough country to ride in, though otherwise an easy place to work because the ranch ran only steers, and the beef rodeo was the only one held during the year.

Bellevue, Lakeside and Poso had good horses but too many gates to open, though the riders ate three meals a day.

When some rider said he was "following" the Miller wagon or the "chain gang" it meant that he rolled up his bed every morning and it was loaded and hauled with the provisions and pots to the next camp where he was to sleep. The last Miller and Lux chuck wagon was put into service in 1912 and was the third and last one under Rafael Cuen in the forty-four years he was vaquero boss for the greatest cattle outfit in the world. With its load of grub, water (sometimes) and bedrolls it was pulled by four little mules which the cook drove. Once located at its camp site, it was the lode star to the far ranging buckeroo, who, having eaten last at four in the morning, would have his thoughts chained to the fire-blackened pots and pans and their contents. He looked to the smoke of the campfire as a shipwrecked sailor looks toward a beacon light on a distant shore.

The cook could have dinner at twelve sharp but in the event the crew was working cattle, even as close as one or two hundred yards from the steaming food, there was no stopping until the work was done, even if it took until five or six in the evening.

Cattle lose weight when held in a herd for any length of time and this must be considered by good vaquero bosses regardless of how hungry the men get. A buckeroo, when asked about the grub, would say, "Well, I never missed any meals when I worked for Miller and Lux, but I have postponed a lot of them."

Headquarters for the wagon and its crew was Corn Camp, some three miles west of Buttonwillow, where the buckeroos had their bunkhouse. From there they traveled in whatever direction they were needed.

The chuck wagon covered an immense territory. It rolled east to the Chester Ranch near the Santa Fe roundhouse on the outskirts of Bakersfield or west to Paso Robles on the coast; north to Alamo Solo and the Fowler in Kings County or southwest to the Cuyama Valley and ranches between points; Carriso Plains, Carneros, Panama, Painted Rock; over the hundreds of miles which made up the southern division of Miller and Lux Ranches.

Cuen said he once counted sixty riders being fed at the wagon, and the crew handled sixty thousand head of cattle. In the twenties, toward its last days as a cattle outfit, the normal crew was fifteen men. I remember a half starved kid so young he still wore knickerbockers, who wandered into Corn Camp and "Papa" Cuen gave him a job. In time he became a salty rider and I suppose he looked on the old vehicle as his home because he followed it until he was a man. This was the case with a lot of kids who learned the vaquero's trade on the Miller and Lux chuckwagon. Some three miles to the north of Buttonwillow stands an enormous old-style barn. It is a land mark which for years has marked the pioneer Tracy Ranch, now run by Cecil, Darryl and Tilton Tracy, and Jack Frey.

Only such a man as Cecil Tracy could have recognized the old chuckwagon's value as a museum piece and as an historical treasure. When the company went out of business and the camp equipment was sold, Cecil, who once followed the wagon, tried to buy it but did not at that time succeed in his efforts. Some years later it changed hands and one day he came upon it attached to a tractor and loaded with hog feed. He entered into a long argument with its owner to rescue the old vehicle from such ignominy and at length prevailed. Now it occupies a place of honor on the ranch, ready to do hospitable service when Cecil has guests for a barbecue. He regrets the "horse shoeing box" is gone but the grub box is still intact.

Though the four little mules which pulled it have long ago gone to the fenceless pastures where the grass is always belly deep, the old wagon, after its many wanderings, has come to rest under a friendly gum tree to wait for time and the elements to wear away this last remnant of a great cattle empire.

Antonio Leivas . . .

Don Antonio Leivas, who could ride as soon as he could walk, was coming from the drive he had made from the San Joaquin Valley to New Mexico. Where Adobe Station road crosses the highway, he found six hundred dollars wrapped in a red bandana. It was very late so he went back to a shack he had passed to make camp. The next morning he found a dead man in the shack. He never knew whether the money belonged to him or not. As he rode away he saw oil seeps on the public land and rode on not knowing what the oil was good for. And so he let fortune pass him by.

The Ghost Horse of Tejon . . .

During rodeo time on Tejon the vaquero was mounted and well on his way before four in the morning. Adolfo, humped with cold, had left camp with the crew long before dawn. He was riding a horse that had a "cold back" and had left his cinch a little loose so that the horse would not get in a bad humor. The riders had reached the mouth of Tunas Canyon when he dismounted to tighten the cinch, the horse having traveled far enough for his back to get warm.

The other men had ridden on and were some little distance when Adolfo mounted and started to overtake them. He had traveled a few yards into the canyon when he heard hoof beats and turned to see a little grey horse trot up beside him. He could make out the animal's outline in the predawn darkness and he saw a stocky, short-coupled little mustang, inclined to roundness with a short wide head and little pin ears; distinctly a Spanish horse of the type that had died out a half century before. It was the most beautiful the vaquero had ever seen. Its tail was

Ghost Horse and *Lancero* Of Tejon

long and its mane hung far down below its neck. Adolfo rode on, the grey trotting along beside him. He put his horse into a gallop and the little horse broke into the same gait keeping pace beside him, its little barb head carried with an arched neck like that of a horse that has been set in the mould of a Spanish bit.

When Adolfo came to the foot of the bridge where the trail climbs out of the canyon the little grey faded away and the vaquero urged his mount up the hill and away from the place of phantom horses.

That night in camp at Alamos, Adolfo told of his meeting with the little mustang. When he had finished, one of the men asked Don Porfirio, the *caporal,* if he believed the lad's story. To the man's surprise the *caporal* said, "Yes, I believe the boy because I too had an experience with a spectral horse one winter night a number of years ago and in the same locality. I once had to make a trip from Tejon to Rancho La Liebre. Circumstances were such that I tried to make it there and back in one day which is not too hard to do. As you know there is a ridge which one can follow from Tunas Canyon across the mountains to Antelope Valley and come out at Pescado and onto the plains a short way from La Liebre. From there the trail leads to Los Angeles. It is an old trail used by honest men and bandits since the first pioneer settlers came from Sonora into California.

"I left headquarters early and riding over the ridge reached my destination by noon. It took some hours to finish my business so it was late in the afternoon before I started on the ride back. The sun was shining in Antelope Valley when I left there but when I had gone about half way, clouds began blowing in from the San Joaquin and I was soon riding in a heavy snow storm.

"I plodded along making slow progress. It became colder and I dismounted and walked to keep from freezing to death. Leading the horse, I soon came to a great oak tree that had fallen; its trunk, being hollow, had broken and split in the fall. There were dry twigs and leaves and having matches and some paper, I soon had a fire going. I crawled into the breached trunk where I fitted snugly and thus with shelter and heat I dozed for some hours. When I woke, the snow had quit falling but a heavy mist had set in over the mountains. As you know, a snowfall changes the aspect of a landscape and I was lost. Knowing the horse would find his way back to his *querencia* I mounted and gave him the rein. After traveling some hours I recognized the long slope that drops into Tunas.

"We had reached the bottom of the canyon when suddenly my horse whirled around with his ears pointing to the back trail, snorting in alarm. Thinking a bear or lion might be following us, I peered into the dense fog. After a time I heard the sound of hoofs slipping and sliding down the mountain. I could tell that a horse and rider were coming as I heard the jingle of spurs and the flapping of tapaderas. I wondered who could be on the ridge on a night like this and if it were some one searching for me, how had he got behind me.

"I called, *'Quién eres tu,* who are you?' No answer came. The hoof beats stopped and I could hear only the sound of the cricket in the bit as the unseen horse spun it with his tongue. I then called again saying, 'Aquí estoy, here I am,' but still no answer. I stared into the fog and soon made out the figure of a horse and its rider who wore a wide brimmed flat crowned sombrero like those one sees in pictures of lanceros who lived in the last century. The horseman wore a beard and seemed to be staring off into the valley. The old style saddle had a *mochila* and long *tapaderos*. The horse was grey and built the same as the one Adolfo has described; a Spanish horse. Suddenly my hair stood on end. I no longer wished to speak with the horseman. I turned my horse around and rapidly rode away. When I reached a bend in the canyon I looked back to see if the *jinete* was following but the rider out of the past had disappeared.

"Some people say the phantom rider on the grey horse was the shade of Murrieta, others, that it was the ghost of Vasquez. Still others say it was the ghost of a young bravo of family who came here to escape the brothers of a man he had killed in a duel.

"Old vaqueros say that old mother Spain sank into impotent decrepitude when her priceless horses degenerated. They swear the phantom greys are the ghosts of faithful mustangs patiently haunting their *querencia,* home pasture, waiting to carry the last vaquero to the final rodeo that he may appear before the great *Mayordomo,* as is fit and proper, with a good horse between his knees.

The talk around the fire then turned to the bandits who had buried gold along old trails in California that lead from Sonoma to far away Sonora and of old Avelino Martinez who, some said had been a bandit or had been in league with them. Others said he had been only a simple sheepherder. Men who really knew said that Avelino had herded sheep in order to keep a large stock of provisions on hand to feed the fugitives without arousing suspicion when they were in hiding near his camp.

Avelino had come to Tejon when California was still Spanish territory and lived to see three flags wave over the state. He was also to see the bandits who kept watch on the peaks pass on and only he remained to sit by the fire and tell stories of another age. Once he had camped in a spot where grizzlies were numerous. He slept on a platform in a tree, his bed held up by four ropes, one on each corner. The bears would visit his camp at night and scatter his pots and pans.

One night an especially noisy visitor was rummaging through his kitchen. Avelino took his gun and leaned over his platform to shoot at the beast. One of the ropes broke and he was spilled on to the back of a badly frightened bear who took off at top speed. The next morning when he was preparing to climb the tree and repair his nest, several men rode up. After grave salutations, one of the riders asked Avelino if they could have meat. He told them to slaughter as many lambs as they needed. He then related his experience with the bear and his visitors set about to help him repair his platform in the tree.

After the lambs had been killed and quartered, the spokesman for the band, in return for Avelino's courtesy and in payment for the sheep, said, "On the other side of this mountain you will find a mesa where three pines stand apart from the other trees and make a triangle. One is blazed. Dig on the side that is blazed and you will find gold. Take what you need." But Avelino never took advantage of the offer. Though he knew or had been told where a number of *entierros,* treasure, were to be found, he never searched for them. He worked for a living and died a poor man. Perhaps, like all *paisanos,* he thought that buried money will be found only by men who have no greed in their hearts and are destined to be rich.

But once, when a vaquero's wife had been ill for a long time and he was desperately in need of money, Avelino heard of his distress and mounting his ancient sorrel horse rode to Tienda del Llano where he found his friend. Being a blunt man, after a brief greeting and without ceremony he said, "*Vale,* I have learned of your bad luck.

I am going to tell you where you can get money but you must not be afraid." He pointed toward the south and continued, "On the side of that mountain is an oak tree which has a branch that curves out and upward like a bent elbow. If you will dig where the point of the elbow aims at the ground you will find a box with money in it; not much but enough for your needs."

The vaquero rode away and found the oak with the branch bent like an elbow. Then he rode back to the ranch store and bought a demijohn of vinegar. Carrying the demijohn and a shovel he rode back to the tree where he dug up a box. He pried the lid off and holding his breath, he poured the vinegar over the money in the box. He did this because buried gold creates a gas which is fatal when breathed, but vinegar counteracts the poisonous fumes. The contents of the box did not exceed a thousand dollars. Much of it was apparently loot from some gambling table or merchandise store, as it was in small coins. It was sufficient for the vaquero's needs as the old man had promised and it was the only *entierros* found on Tejon of the many that legend says were buried there in the old days.

Camp Fire Tales . . .

The treasure of Tunas Canyon has never been found, though the gold pieces which filled two buckskin bags and made a load for a pack horse lies deep in a fissure of a cliff at the mouth of a *barranca* beside the trail traveled by many men for a century or more.

Once the red men had lived here in abundance and drove great herds of elk up the canyon to be killed by hunters hidden in the hills to provide meat for many days of feasting. But lean days had come to the Indian and hunger stalked the camp in Tunas when the bandit, the girl and the pack horse load of gold he had stolen, came riding up the creek and was captured by the young men and brought before the chief who spared his life because he had the girl with him and warriors do not war on women. The bandit rode away with his companion, unharmed, but the pack horse was relieved of his burden only to be killed for food.

In those days chiefs had power and authority and their orders were obeyed without question. Knowing full well that the misery and wretchedness which was the Indians' lot was caused by lust for gold, the chief ordered the bandit's loot to be buried deep under ground so that the clean earth would cover the glitter which lit the fires of greed and cupidity in the heart of the white man and brought suffering, outrage and death to the redman.

In time, the chief and his people went the way of all Indians but one of the young men who had hidden the gold, when old and near his death, told enough about the bandit, the gold and the lady to leave behind him a legend of buried treasure and so provide a subject of conversation and conjecture around the campfire and stir a wistful, "*Ojala,* if God would will, I could find it" from some vaquero lying in his bed on the hard ground and dreaming of an easier, fuller life.

The ashes from the campfire on Alamos Creek have been scattered to the four winds these many years, and the men who were warmed by its flames have all ridden over the great divide. Only Adolfo remains, who told me the story of a night around the fire when each man recounted some tale or told of something that had happened to him in the years spent riding the Tejon range. The vaquero, often unable to read or write, and who was never more than a herder of cattle was, nevertheless, a matchless story teller. Flickering flames never failed to light his talent for unfolding a tale.

Evening is the pleasantest part of the range riders day for then the *platica,* a word which has no equal in the English language and means cheerful, courteous conversation, turns to various incidents of the day and ends up with stories of brave horses, bandits and buried treasure.

That night as is usual around the campfire during rodeo time, the men gathered to sit and smoke. Don Antonio Leivas, who had started to ride as soon as he could walk, and being the oldest in the crew, began the evening's conversation with a tale about the Barileño mustang and how he came to run wild in Antelope Valley.

This wild horse had its origin in the thirty-five mares and three stallions of purest Barbary strain, brought into the desert and turned out to range at Barilito Springs by the ancestors of Don Jesús Lopez. The stallions fought for many days before the strongest conquered and killed his rivals. Thus the fittest survived to sire the wiry, compact mustang famous the length of California for endurance; that one quality demanded in a vaquero's horse for which all vices are forgiven. The high desert valley where conditions closely approached those of their African ancestry was ideal for the horses and they throve and multiplied. Nature provided them with a color to blend with the landscape; buckskin, *grullos,* crane colored, and golden palomino.

The Tejon riders captured many bands and selecting the best, bred their saddle stock from these wild mares. The many fugitives from justice or vengeance who found sanctuary in the sierra were thankful for their staying qualities when running from the law. Don Antonio, the *caporal,* for thirty years drove bands of them to San Francisco and as far east as Albuquerque through the

The Palomino Horse (second from left)
Is a Pure Blooded Barileño Mustang (from a 1901 photo)

hunting grounds of Indian tribes and never lost a horse to any of them. Starting with the Mojaves, who helped him across the Colorado and into the territory of the next tribe, he and his vaqueros with the band of horses passed from Mojaves to Utes to Navajos and to their final destination in New Mexico.

Don Antonio witnessed the death of the last wild stallion in a noose set to trap him. The palomino had been pursued by mounted men for many days before he turned into Sacatera Canyon where he ran into the loop hanging from a stout sapling. He died there fighting the rope which strangled him and ended the finest strain of horses in California.

In later years Don Jesús Lopez, the mayordomo, when sore pressed for horse flesh, would sigh for the tireless mustangs of his youth and the qualities he could not find in the Tejon horses after the old Spanish breed had passed out of existence. "Ah," he would say, "these horses are not like the ones of the old days; these have no stamina or endurance."

Adobe Houses . . .

There are ruins of two adobe houses near the spot where Padre Garces forded Kern River. John Barker, according to Canducci Castro, was the first man to come into that part of the San Joaquin. Barker said that the two adobe houses were there even then where the bridge crosses the river on the Olcese Ranch.

Who built them nobody knows. It could be that they were attempts at settlement and were abandoned like the old ranches which skirt the great valley such as Pleito, Temblor and Carneros.

Expert Stockman . . .

The old-timer was broke and out of a job. One by one the great ranches had passed out of existence and no one needed the cowboy who had grown gray riding the ranges, a pathetic figure left along the path of progress, with no place to apply his hard won skill. His possessions were an old saddle, a few tools and a battered car. The old man's face was weather beaten and his eyes had that squint which is caused by years of staring across sunlit spaces, but his back was still straight as a ramrod and his face still resolute and without fear. He knew that scattered across the valley were farms where a horse or two were kept for pleasure, or a bull or steer for beef; he would only have to find the ones that needed work done on them. He started the old car and drove along the highway as he had done many other mornings, keeping a sharp eye out for horses on the farms.

He had traveled about an hour when he saw, over to his right, a roan mare and a yearling Holstein bull in a wire corral near a farm house. He drove into the farmyard and was looking the mare and bull over when the farmer approached him. The mare's feet were in bad shape and the cowboy asked, "Do you want the mare's feet trimmed and the bull altered?" The farmer answered, "Yes, can you do it?" The old cowboy assured him that he could and proceeded to pull his saddle, bridle and tools out of the car. He turned to the farmer and asked, "Does the mare ride?"

The farmer replied, "I don't know. She was here when I bought the place three years ago. My kids are afraid to try and ride her. I have never seen her ridden."

The cowboy caught the mare and trimmed her feet, then he saddled her. He wrapped the reins around the horn and led her around in a circle. After a turn or two, he got her into a gallop. The roan made a jump, then lowered her head to buck but the bit dissuaded her; she raised her head, galloped a little distance, but the bit again discouraged her and she gave up trying. The vaquero knew that he could hold her head up without her falling over backward, so he mounted and rode her out across a field until she got the hump out of her back and she worked her ears back and forth. He rode back and put the farmer's son on the mare and let him ride for a while.

The bull was in the wire corral from which he wandered in and out at will. The vaquero mounted the mare, and taking the rope down, maneuvered the bull beside a hay stack where he could throw his rope. The gentle bull walked along but started to run when the cowboy whirled his rope and threw the loop around the stubs of horns. The little roan mare turned out to be a first class stock horse, for when the bull was on the ground with a rope around his hind legs and the cowboy was working on him, the mare braced her little legs and held the rope tight until the vaquero had finished his task.

The farmer gave the old cowboy eight dollars and was well satisfied; the mare could walk, his children could now ride her and the calf would now make beef. The old man loaded his saddle and tools and drove away to look for another odd job. The farmer, watching him drive away, regretted that a man of so much skill and knowledge should be brought to such a hand-to-mouth existence.

This And That . . .

I have forgotten my source of information but I offer the following for what it is worth:

Mountain lions would slyly get between a colt and the rest of the herd then drive the colt away and kill it. A lion could drag a large colt for a long distance until he found a place to bury the carcass with leaves and brush.

Quite often a single lion dog would tree a lion. Lion dogs were made fearless by feeding them the lion's meat; the surest way of attaining this end was to take a quantity of clotted blood and pour it down the dogs throat, making him swallow it.

A portion of the lion's grease or tallow was also saved, particularly by those who owned race horses. Before a race the horse's rump and tail were smeared with this grease. Besides scaring the horse into running faster to get away from the odor, the other horses would not pass the greased one for fear of the lion scent; but on the other hand, if the horse without the lion grease got a head start, he would run because the horse carrying the lion scent would scare him into staying ahead of the lion.

Wine smeared on the horses in the same places had the same effect but in a lesser degree.

When a rattlesnake was killed the gall was always extracted as this applied to a snake bite would draw the venom from the wound. It is said that a rattlesnake will not bite a pregnant woman, no matter how close to the snake she gets. Also, if a pregnant woman spits on a rattlesnake it will die. Another old belief is that a rattler removes its poison sac when it takes a drink of water and if something should happen to the sac and it should be carried awav or lost, the snake will kill itself.

The End

Wild Hog (see pg. 120)

Doane

Made in United States
Orlando, FL
04 November 2026

90220725R00099